Advances in Intelligent Systems and Computing

Volume 727

Series editor

Janusz Kacprzyk, Polish Academy of Sciences, Warsaw, Poland
e-mail: kacprzyk@ibspan.waw.pl

The series "Advances in Intelligent Systems and Computing" contains publications on theory, applications, and design methods of Intelligent Systems and Intelligent Computing. Virtually all disciplines such as engineering, natural sciences, computer and information science, ICT, economics, business, e-commerce, environment, healthcare, life science are covered. The list of topics spans all the areas of modern intelligent systems and computing such as: computational intelligence, soft computing including neural networks, fuzzy systems, evolutionary computing and the fusion of these paradigms, social intelligence, ambient intelligence, computational neuroscience, artificial life, virtual worlds and society, cognitive science and systems, Perception and Vision, DNA and immune based systems, self-organizing and adaptive systems, e-Learning and teaching, human-centered and human-centric computing, recommender systems, intelligent control, robotics and mechatronics including human-machine teaming, knowledge-based paradigms, learning paradigms, machine ethics, intelligent data analysis, knowledge management, intelligent agents, intelligent decision making and support, intelligent network security, trust management, interactive entertainment, Web intelligence and multimedia.

The publications within "Advances in Intelligent Systems and Computing" are primarily proceedings of important conferences, symposia and congresses. They cover significant recent developments in the field, both of a foundational and applicable character. An important characteristic feature of the series is the short publication time and world-wide distribution. This permits a rapid and broad dissemination of research results.

More information about this series at http://www.springer.com/series/11156

Siddhartha Bhattacharyya
Anirban Mukherjee · Hrishikesh Bhaumik
Swagatam Das · Kaori Yoshida
Editors

Recent Trends in Signal and Image Processing

ISSIP 2017

Springer

Editors
Siddhartha Bhattacharyya
Department of Computer Application
RCC Institute of Information Technology
Kolkata, West Bengal
India

Anirban Mukherjee
Department of Engineering Science and Management
RCC Institute of Information Technology
Kolkata, West Bengal
India

Hrishikesh Bhaumik
Department of Information Technology
RCC Institute of Information Technology
Kolkata, West Bengal
India

Swagatam Das
Electronics and Communication Sciences Unit
Indian Statistical Institute
Kolkata, West Bengal
India

Kaori Yoshida
Department of Human Intelligence Systems
Kyushu Institute of Technology
Wakamatsu-ku, Kitakyushu, Fukuoka
Japan

ISSN 2194-5357 ISSN 2194-5365 (electronic)
Advances in Intelligent Systems and Computing
ISBN 978-981-10-8862-9 ISBN 978-981-10-8863-6 (eBook)
https://doi.org/10.1007/978-981-10-8863-6

Library of Congress Control Number: 2018935214

Printed on acid-free paper

This Springer imprint is published by the registered company Springer Nature Singapore Pte Ltd. part of Springer Nature
The registered company address is: 152 Beach Road, #21-01/04 Gateway East, Singapore 189721, Singapore

Siddhartha Bhattacharyya would like to dedicate this book to his father Late Ajit Kumar Bhattacharyya, his mother Late Hashi Bhattacharyya, his beloved and evergreen wife Rashni, his cousin brothers Prithwish, Santi, Smritish, Palash, Pinaki, Kartick and Atanu

Anirban Mukherjee would like to dedicate this book to Late Mr. P. K. Sen, former Head, Department of IT, RCCIIT

Hrishikesh Bhaumik would like to dedicate this book to his late father, Major Ranjit Kumar Bhaumik, his greatest inspiration and to his mother Mrs. Anjali Bhaumik who has supported and stood by him in all ups and downs of life

Swagatam Das would like to dedicate this book to his beloved wife Sangita Sarkar

Kaori Yoshida would like to dedicate this book to everyone who is passionate to image and signal processing research

Preface

In this era of technology, almost every modern tools and gadgets resort to signal processing in one way or the other. The algorithms governing mobile communications, medical imaging, gaming, and host of other technologies all encompass some kind of signal processing. The signals might be speech, audio, images, video, sensor data, telemetry, electrocardiograms, or seismic data among others. The possible application areas include transmission, display, storage, interpretation, classification, segmentation, or diagnosis. The signals generally handled in real-life situations are often uncertain and imprecise, often posing a challenge and requiring advanced computational techniques to process and analyze. Scientists and researchers all over the world are extensively investing efforts in developing time-efficient, robust, and fail-safe signal processing algorithms for the benefit of mankind.

2017 First International Symposium on Signal and Image Processing (ISSIP 2017) organized at Kolkata during November 01–02, 2017, was aimed to bring together researchers and scholars working in the field of signal and image processing. This is a quite focused domain yet broad enough to accommodate a wide spectrum of relevant research work having potential impact. The symposium showcased author presentations of 21 high-quality research papers carefully selected through a process of rigorous review by experts in the field. In the present treatise, all these 21 research papers have been meticulously checked and compiled with all necessary details following the Springer manuscript guidelines. It is indeed encouraging for the editors to bring out this collection under the Springer book series of Advances in Intelligent Systems and Computing (AISC). The organization of this book containing 21 papers as separate chapters is as follows:

A novel hybrid algorithm is presented by the authors of Chapter "Design of Higher Order Quadrature Mirror Filter Bank Using Simulated Annealing-Based Multi-swarm Cooperative Particle Swarm Optimization" for obtaining prototype filter that leads to near-perfect reconstruction for lower- and higher-dimensional filter banks. A comparison of the algorithm made with other existing methods reveals a significant increase in stop-band attenuation and reduction in perfect reconstruction error (PRE) of 82-tap filter bank.

Chapter "Medav Filter—Filter for Removal of Image Noise with the Combination of Median and Average Filters" also deals with a hybrid filter for removal of image noise. It is better than the primitive filters in terms of edge preservation and signal-to-noise ratio (SNR) when the intensity of disrupted noise is very high.

A neural network-based classifier is proposed in Chapter "Classification of Metamaterial-Based Defected Photonic Crystal Structure from Band-Pass Filter Characteristics Using Soft Computing Techniques" that deals with the classification problem of metamaterial-based photonic crystal from its band-pass filter characteristics. High accuracy of the classifier must attract the attention of the researchers.

Chapter "Sparse Encoding Algorithm for Real-Time ECG Compression" deals with encoding algorithm of ECG signals. Here, the authors propose and validate a sparse encoding algorithm consisting of two schemes, namely geometry-based method (GBM) and the wavelet transform-based iterative thresholding (WTIT).

The authors of Chapter "Wavelet Based Fractal Analysis of Solar Wind Speed Signal" have studied the presence of multi-fractality of solar wind speed signal (SWS). Wavelet-based fractal analysis has been employed for this purpose, and qualitative evaluation is also shown.

Clinical importance of electromyogram (EMG) signals is immense for diagnosis of neuromuscular diseases like neuropathy and myopathy. Authors in Chapter "Class Discriminator-Based EMG Classification Approach for Detection of Neuromuscular Diseases Using Discriminator-Dependent Decision Rule (D3R) Approach" have demonstrated a new method of classification of EMG signals, based on SVM, which can be reliably implemented in clinical environment.

Real-life signal and image processing applications often entail medium- to large-scale multi-objective and many-objective optimization problems involving more than hundred decision variables. Chapter "A Cooperative Co-evolutionary Approach for Multi-objective Optimization" proposes an evolutionary algorithm (EA) that can handle such real-world optimization problem with reasonable accuracy.

Vehicle tracking through smart visual surveillance is an important part of intelligent traffic monitoring system that is gaining wider application day by day. Authors in Chapter "Automatic License Plate Recognition" have addressed this important practical issue by proposing a novel technique of automated license plate recognition of moving vehicles. They have worked on two different databases of traffic video justifying impressive performance of their proposed technique primarily with respect to recognition accuracy.

Quality of classification depends on accuracy of selection of prominent features after removing irrelevant and redundant data from a high-dimensional data set. The authors of Chapter "S-shaped Binary Whale Optimization Algorithm for Feature Selection" have proposed and evaluated an effective algorithm for finding optimal feature subset from a given data set.

Chapter "Motion Artifact Reduction from Finger Photoplethysmogram Using Discrete Wavelet Transform" deals with noise reduction from photoplethysmogram (PPG) signal obtained at fingertip. Motion artifact is injected into the clean PPG signal artificially, and denoising is done using discrete wavelet transform. Comparative analysis shows that the performance of the proposed method is better than the existing ones.

Precision of automatic target recognition and striking has become an important area of modern defense research and development. Real-time target classification and recognition require real-time processing of high-frequency and higher-precision THz signals over an ultra-wide bandwidth. Authors in Chapter "Analysis of Picosecond Pulse for ATR Using Ultra-Wideband RADAR" have taken up this very sensitive work of spectrum analysis of THz pulses for detecting radar target.

Authors in Chapter "Detection of Endangered Gangetic Dolphins from Underwater Videos Using Improved Hybrid Frame Detection Technique in Combination with LBP-SVM Classifier" have taken up a very interesting problem of detecting aquatic organisms like fish and dolphins in underwater poor lighting condition. Underwater video is analyzed and processed to recognize endangered Gangetic dolphin class using the hybrid of traditional SVM classifier and local binary pattern feature extractor.

Lip contour detection and extraction is the most important criterion for speech recognition. Chapter "Automatic Lip Extraction Using DHT and Active Contour" presents a new lip extraction algorithm that works good in case of uneven illumination, effects of teeth and tongue, rotation, and deformation.

Noise classification is very crucial in medical image processing mainly because of the associated medical implication. Chapter "Early Started Hybrid Denoising Technique for Medical Images" deals with a hybrid denoising technique for brain images obtained by PET and CT scans, and authors share some of their impressive findings in this regard.

Chapter "Intelligent Tutoring by Diagram Recognition" demonstrates a nice application of digital diagram recognition and analysis in facilitating student's learning of geometry. The authors have reported a case study of elementary geometry of primary school level and have shown how intelligent handling of digital image can replace traditional teaching.

Quantum computing is a new paradigm of intelligent computing. Authors in Chapter "Color MRI Image Segmentation Using Quantum-Inspired Modified Genetic Algorithm-Based FCM" have deployed quantum-inspired modified genetic algorithm for color MRI image segmentation that has enhanced the speed, optimality, and cost-effectiveness of the conventional GA or modified GA.

Processing and digitization of handwritten documents is an important application of clustering algorithms. Chapter "Multi-verse Optimization Clustering Algorithm for Binarization of Handwritten Documents" presents an automatic clustering algorithm for binarization of handwritten documents based on multi-verse optimization. The proposed approach is tested on a benchmark data set.

Effectiveness of 3D object reconstruction and recognition from a set of images is evaluated in Chapter "3D Object Recognition Based on Data Fusion at Feature Level via Principal Component Analysis." Different feature extraction, matching, and fusion techniques and discrete wavelet transform are used to reconstruct different 3D models from a given set of images.

With newer techniques evolving for signal and image processing, unauthorized manipulation and corruption of digital audio, image, and video data is becoming easier, thereby requiring robust watermarking technique. Authors have offered a new watermarking technique for digital image (for copyright protection) using discrete wavelet transform (DWT) and encryption in Chapter "Digital Image Watermarking Through Encryption and DWT for Copyright Protection."

Extraction of textural and acoustic features from speech and non-speech audio files and classification of audio files comes under the purview of Chapter "Speech and Non-speech Audio Files Discrimination Extracting Textural and Acoustic Features." This is a new interesting area of research of audio signal recognition.

Another interesting area of audio signal recognition is speech recognition. Chapter "Speaker Recognition Using Occurrence Pattern of Speech Signal," the last chapter, addresses speaker identification problem that has potential application in forensic science, tele-banking, smart devices, etc. Authors have shown how their method correctly classifies speech sample and identifies the speaker.

This treatise contains 21 chapters encompassing various applications in the domain of signal and image processing. The applications range from filtering, encoding, classification, segmentation, clustering, feature extraction, denoising, watermarking, object recognition, reconstruction, fractal analysis on a wide range of signals including image, video, speech, non-speech audio, handwritten text, geometric diagram, ECG and EMG signals, MRI, PET, and CT scan images, THz signals, solar wind speed signals (SWS), and photoplethysmogram (PPG) signals. The authors of different chapters share some of their latest findings that can be considered as novel contributions in the current domain. It is needless to mention that the effort by the editors to come out with this volume would not have been successful without the valuable contribution and the effort and cooperation rendered by the authors. The editors also would like to take this opportunity to express their thanks to Springer as an international publishing house of eminence to provide the scope to bring out such a concise and quality volume. The editors would also like to express their heartfelt thanks to Mr. Aninda Bose, Senior Editor, Springer, for his support and guidance right from the planning phase. The editors also express their gratitude to the respected reviewers who have shared their valuable time and expertise in meticulously reviewing the papers submitted to the symposium and finally selecting the best ones that are included in this volume. We sincerely hope that this book volume becomes really useful to the young researchers, academicians, and scientists working in the domain of signal and image processing and also to the postgraduate students of computer science and information technology.

Kolkata, India — Siddhartha Bhattacharyya
Kolkata, India — Anirban Mukherjee
Kolkata, India — Hrishikesh Bhaumik
Kolkata, India — Swagatam Das
Kitakyushu, Japan — Kaori Yoshida

Contents

About the Editors

Siddhartha Bhattacharyya, D. Litt. [FIEI, FIETE, LFOSI, SMIEEE, SMACM, SMIETI, LMCSI, LMISTE, LMIUPRAI, MIET (UK), MIAENG, MIRSS, MIAASSE, MCSTA, MIDES, MISSIP] did his BS, B.Tech., and M.Tech. from the University of Calcutta in 1995, 1998, and 2000, respectively. He completed his Ph. D. (Engg.) from Jadavpur University in 2008. He is the recipient of several awards like *Distinguished HoD Award* and *Distinguished Professor Award* (both 2017) and *ICT Educator of the Year* (2017). He is a *Distinguished Speaker of ACM*, USA, for the period 2018–2020. He is currently the Principal of RCC Institute of Information Technology, Kolkata, India. He is a co-author of 4 books and co-editor of 10 books, and has more than 190 research publications.

Anirban Mukherjee is currently Associate Professor in the Department of Engineering Science and Management at RCC Institute of Information Technology, Kolkata, India. He did his BE from Jadavpur University in 1994 and PDOM from IGNOU, New Delhi, in 1998. He completed his Ph.D. (Engg.) from IIEST, Shibpur, in 2014. He has over 18 years of teaching and 10 years of research experience. He has published 4 journal papers, 2 chapters, 9 conference publications, and 19 books (including 2 textbooks for UG engineering courses and 16 co-authored books for 6 universities in India on Computer Graphics & Multimedia).

Hrishikesh Bhaumik is currently serving as an Associate Professor in the Department of Information Technology at RCCIIT. He did his BS from the University of Calcutta in 1997, AMIE in 2000, and M.Tech. from BE College, Shibpur, in 2004. He is with RCCIIT for more than 13 years. He received several sponsorships from ICTP, Italy, and acted as a major contributor to the EU-India Grid Project during 2010 to 2012. He has acted as a coordinator and resource person in several AICTE/ISTE sponsored workshops held in different parts of India. He was previously the Head of Department of IT at RCCIIT.

Swagatam Das is currently an Assistant Professor at ECSU of Indian Statistical Institute, Kolkata. He has published one research monograph, one edited volume, and more than 150 research articles in peer-reviewed journals and international conferences. He is the founding co-editor-in-chief of Elsevier journal Swarm and Evolutionary Computation. He serves as associate editors of IEEE Trans. on Systems, Man, and Cybernetics: Systems, IEEE Computational Intelligence Magazine, IEEE Access, Neurocomputing, Information Sciences, and Engineering Applications of Artificial Intelligence. He is the recipient of the 2012 Young Engineer Award from the Indian National Academy of Engineering (INAE). He has 7500+ Google Scholar citations and an H-index of 44 till date.

Kaori Yoshida studied Electrics and Computer Engineering at Kyushu Institute of Technology and received her bachelor and master's degrees. From 1996 to 1999, she has worked as intern at Electrotechnical Laboratory, AIST, MITI in Tsukuba City, Japan. She has obtained JSPS Research Fellowship from 1998 and achieved doctoral degree at Kyushu Institute of Technology in 1999. She is currently an Associate Professor in the Department of Brain Science and Engineering at Kyushu Institute of Technology. Her research interests include Kansei Information Processing, Human–Computer Interaction, and Soft Computing. She has published more than 100 research papers in these fields.

Design of Higher Order Quadrature Mirror Filter Bank Using Simulated Annealing-Based Multi-swarm Cooperative Particle Swarm Optimization

Supriya Dhabal, Roshni Chakrabarti and Palaniandavar Venkateswaran

Abstract This paper presents a novel hybrid algorithm based on Multi-swarm Cooperative Particle Swarm Optimization (MCPSO) and Simulated Annealing (SA) for the design of higher order Quadrature Mirror Filter (QMF) bank. The optimization of lower order filters can be carried out easily by traditional optimization methods, but these approaches failed to find higher order filter coefficients due to nonlinear and multimodality problem space. Most of the optimization algorithms are easily trapped into local optimum which yields few unwanted characteristics in filter magnitude responses like ripples in transition region, lower stop-band attenuation. The proposed algorithm, named Simulated Annealing-based Multi-swarm Cooperative PSO (SAMCPSO), is presented here to obtain prototype filter that leads to near-perfect reconstruction for both lower and higher dimensional filter banks. Comparison with other existing methods in the literature demonstrates that the proposed algorithm exhibits an average increase of 17.39% in stop-band attenuation and 47.35% reduction in Perfect Reconstruction Error (PRE) of 82-tap filter bank.

Keywords QMF · Filter bank · NPR · PSO · Metropolis criterion · SAPSO

S. Dhabal (✉)
Netaji Subhash Engineering College, Kolkata, India
e-mail: supriya_dhabal@yahoo.co.in

R. Chakrabarti
Techno India Banipur, Habra, West Bengal, India
e-mail: roshni.078@gmail.com

P. Venkateswaran
Jadavpur University, Kolkata, India
e-mail: pvwn@ieee.org

S. Bhattacharyya et al. (eds.), *Recent Trends in Signal and Image Processing*,
Advances in Intelligent Systems and Computing 727,
https://doi.org/10.1007/978-981-10-8863-6_1

1 Introduction

During the last few years, QMF bank has been widely employed in the processing of biomedical signals, design of wavelet bases, noise cancellation, discrete multi-tone modulation systems, wideband beam-forming, and so on [1–5]. Few applications of filter bank like echo cancellation, cross talk suppression, ECG signal processing, three-dimensional audio reduction systems, and efficient realization of the higher order filter banks are necessary as they necessitate high attenuations at stop-band. Therefore, several design approaches are developed for efficient realization of the filter bank. Due to the complex optimization and high degree of nonlinearity, the techniques provided in [1–3, 6–12] are not applicable for the design of higher order filter banks and do not have precise control at transition band of the magnitude response which yields in sub-optimal solutions. The above-mentioned limitations can be tackled effectively by integrating SA with MCPSO.

Although PSO is considered as a robust algorithm exhibiting fast convergence in many practical applications, it suffers from the premature convergence problem and can be easily trapped in local optima. Therefore, several attempts have been made by researchers to improve the performance of PSO. One such method is the hybridization of PSO with other local search methods like SA. By hybridizing SA with PSO, significant performance improvement is achieved for different practical problems. Besides this, the recent development in optimization paradigm shows that the multi-swarm-based methods can perform better with the higher dimensional and multimodal problems [13–17]. Thus, by exploiting improved features of SA and multi-swarm PSO, a new SAMCPSO algorithm has been proposed in this paper towards the efficient design of higher order QMF bank.

The remaining of the paper is arranged as follows: the formulation of design problem to obtain prototype filter is discussed in Sect. 2. The proposed design method using hybrid SAMCPSO is presented in Sect. 3. The simulation results with different design examples are demonstrated in Sect. 4. Further, comparative study of the performance of QMF bank with other existing methods is also analysed and presented. Finally, conclusions and further research scopes are given in Sect. 5.

2 Design Problem Formulations

QMF is a two-channel filter bank, consisting of analysis filters followed by down-samplers in the transmitting part and up-samplers with synthesis filters in the receiver section, as presented in Fig. 1. The design problem of QMF can be solved by considering mean square errors in passband (E_p), stop-band (E_s), transition band (E_t), and ripple energy of filter bank, i.e. Measure of Ripple (MOR) [7]. During the design of filter bank, it is assumed that the prototype filter should be ideal in passband and stop-band, and there exists lowest possible reconstruction error at

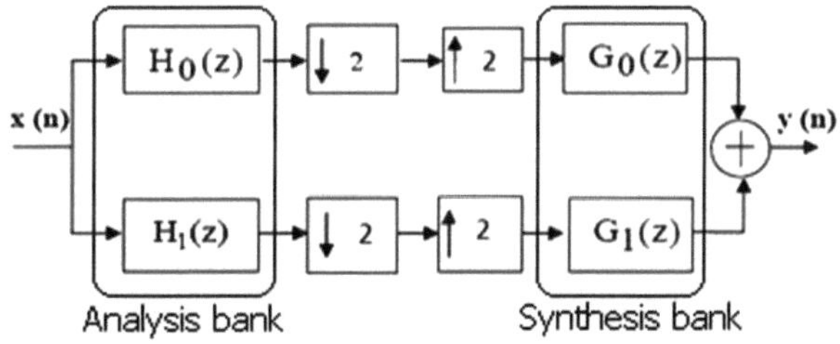

Fig. 1 Basic structure of QMF bank

$\omega = \pi/2$. Consequently, the objective function (φ) is formulated using a weighted sum of these four parameters as given by [2–7]:

$$\varphi = \alpha E_p + (1 - \alpha) E_s + E_t + \beta\, \mathrm{MOR} \tag{1}$$

where $0 < \alpha \leq 1$ and β are the weight of MOR.

3 The Proposed Hybrid SAMCPSO Algorithm

Multi-swarm cooperative PSO (MCPSO) is an important variant of the basic PSO algorithm, based on the use of multiple swarms instead of a single swarm [14–17]. In MCPSO method, the whole population is divided into a number of sub-swarms which can execute independently following the principle of master–slave model: it consists of one master swarm and several slave swarms. Each slave swarm executes a single PSO or its variants. When all the slave swarms are equipped with the new solutions, each of them sends the best local solution to the master. Thereafter, master swarm selects the best of all received solutions and evolves velocity and position according to the following modified equations:

$$\begin{aligned} V_{iD}^{M} = \chi \Big(& \omega * V_{iD}^{M} + c_1 * \frac{r_1}{r} * (pbest_{iD}^{M} - X_{iD}^{M}) \\ & + \psi * c_2 * \frac{r_2}{r} * \left(pbest_{gD} - X_{iD}^{M}\right) \\ & + \psi * c_3 * \frac{r_3}{r} * \left(pbest_{gD}^{M} - X_{iD}^{M}\right) \\ & + (1 - \psi) * c_4 * \frac{r_4}{r} * (pbest_{gD}^{S} - X_{iD}^{M}) \Big) \end{aligned} \tag{2}$$

$$X_{iD}^{M} = X_{iD}^{M} + V_{iD}^{M} \tag{3}$$

where $c_{1-4} = 2.05$, $r_{1-4} \in [0, 1]$ are random numbers related as $r = r_1 + r_2 + r_3 + r_4$, and ψ is a migration factor given by

$$\psi = \begin{cases} 0 \ pbest_g^S \leq pbest_g^M \\ 1 \ pbest_g^S > pbest_g^M \end{cases} \tag{4}$$

Here, ψ is responsible for supplying best slave swarm to the master swarm as the generation proceeds. The best performed slave and master particles are represented by $pbest_g^S$ and $pbest_g^M$, respectively. As inertia weight (ω) plays a significant role in balancing act of exploration and exploitation behaviour of particles, a new adaptive inertia weight is introduced here for the adjustment of ω as:

$$\omega = \begin{cases} \omega\text{max} & \text{if} \quad f \geq f_{avg} \\ \omega\text{max} - (f - f_{min}) * (\omega\text{max} - \omega_{min})/(f_{avg}/4 - f_{min}) & \text{if } f \geq f_{avg}/4 \,\&\, f < f_{avg} \\ (\omega\text{max} - \omega_{min}) \times \frac{(\text{iter}_{max} - \text{iter})}{\text{iter}_{max}} + \omega_{min} & \text{if} \quad f < f_{avg}/4 \end{cases} \tag{5}$$

where "f" denotes the fitness of current particle, f_{avg} is the mean fitness of the swarm selected, and f_{min} indicates the fitness of global best solution achieved so far. In our proposed approach, two modifications are performed to escape from local minima while maintaining the diversity of particles in the swarm: (a) the SA is introduced inside the search process because SA employs high probability to jump out from local optima [10, 13], and (b) instead of single swarm, the cooperative multi-swarm scheme with modified search equation and inertia weight is employed which helps in better balancing the exploration and exploitation performance of particles in the swarm [15]. Thus, the effective combination of SA and multi-swarm cooperative PSO scheme reduces the computational complexity for searching the lower order filter coefficients, and at the same time, it also has the sufficient ability to avoid premature convergence for the design of higher order filters. The summary of proposed SAMCPSO is introduced as follows:

Step 1: **Initialization stage**—Specify number of filter taps, the passband frequency (ω_p), stop-band frequency (ω_s), α, and β . **PSO**: Initialize number of slave swarms and size of each swarm; create initial swarms by randomly generating position and velocity vectors for each particle assuming first (*MM*) entries are zero; initialize ω_{max}, ω_{min}, c_1, c_2, maxNOI (maximum number of iterations) and NOI = 0, ite_M = no. of iterations after which MM will be reduced. **SA**: Assume *LS*—local search interval, initial temperature (T_0) , minimum temperature $\left(T_{min}\right)$. For simplification purpose, cooling schedule for SA is chosen as $T\,(k) = 0.99 * T(k-1)$.

Step 2: **Computation for PSO and SA**—Evaluate the fitness of all particles in each swarm; find out $pbest_g^s$ from each slave swarm, $pbest_g^M$ from master swarm and $gbest$ for the entire swarm. Repeat until evaluations <maxNOI or $T\,(k) < T_{min;}$ select a random particle from each swarm and update velocity and position of slave and master particles. Calculate fitness of newly generated slave and master particles. Update best slave $pbest_g^s$ from each group and best master $pbest_g^M$. Update $gbest$ based on Metropolis criterion [13]. If $gbest$ is modified, then reduce temperature using the cooling schedule. After every *LS* evaluation, perform SA for $gbest$ particle.

If $MM \neq 0$ and mod (iterations, ite_M) = 0, then reduce $MM = MM - 1$ and NOI = NOI + 1.

Step 3: **Output optimization results**—Design the prototype filter using best solution of the whole swarm, i.e.*gbest*, and obtain all other filters of QMF bank from the prototype filter.

4 Simulation Results and Discussion

In this section, the proposed SAMCPSO method is used for the design of prototype filter of QMF bank. The unknown parameters of cost function, i.e. α and β are selected by trial and error methods to obtain the best possible solutions. The complete simulation work is carried out using MATLAB 2012a on Genuine Intel(R) Core (TM) i5-2450 M @2.5 GHz, 4 GB RAM. In the design, following performance parameters are measured: square errors in passband (E_p), stop-band (E_s), transition band (E_t),

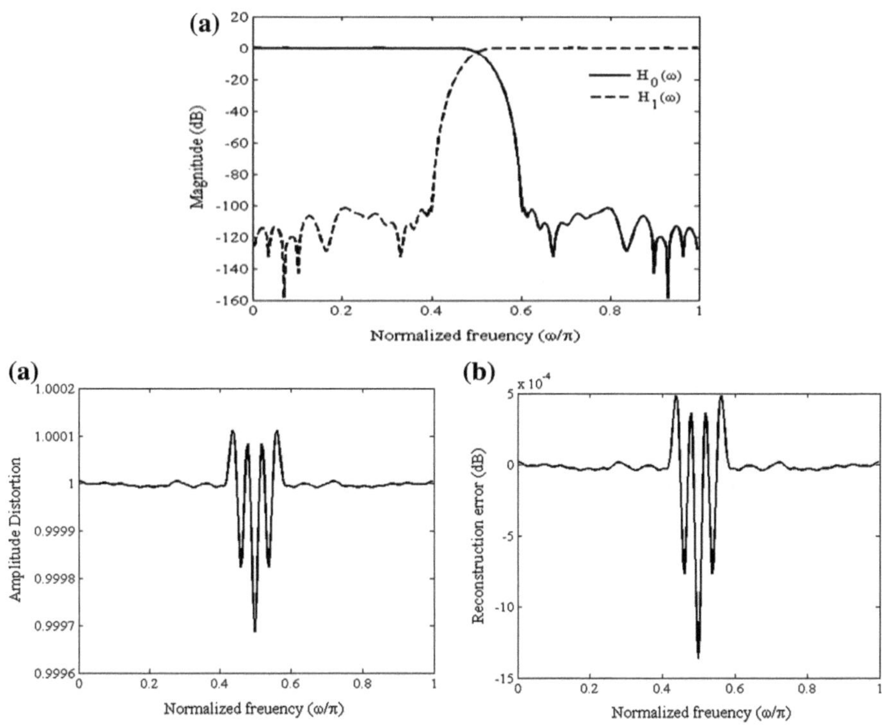

Fig. 2 Frequency response of QMF for N = 100: **a** amplitude response of analysis filters, **b** variation of amplitude distortion, **c** reconstruction error in dB

overall amplitude distortion $e_{am}(\omega)$, Peak Reconstruction Error (PRE), and stop-band attenuation (A_s) in dB [7].

4.1 Design Problem

Here, the filter bank is designed with the set of specifications: N = 100, $\omega_p = 0.4\pi$, and $\omega_s = 0.6\pi$. The magnitude response for analysis filters and amplitude distortion function are plotted and shown in Fig. 2a and Fig. 2b, respectively. Figure 2c illustrates the reconstruction error of QMF bank. The calculated values of A_s and PRE are 122.23 dB and 0.000334 dB, respectively. The remaining performance results are as follows: $E_p = 2.94 \times 10^{-12}$, $E_s = 8.57 \times 10^{-14}$, and $E_t = 2.65 \times 10^{-19}$.

4.2 Performance Analysis for Designing Higher Order Filter Banks

For higher values of N, the performance results are shown in Table 1. The values of α, and β are selected based on the minimum of fitness function achieved. It is obvious that with the increase of filter taps (N), the values of E_p, E_s, E_t, e_{am}, and PRE are reduced gradually while A_s is increased.

4.3 Comparison of Results with Other Algorithms

Table 2 indicates the improvement in performance for higher order filter banks in comparison with recently available methods given in [4, 5, 11, 18–20]. For N = 48, keeping E_p, E_s, and E_t almost at the same level, the percentage improvements in A_s are 17.27%, 15.38%, 7.63%, 23.04%, 2.68%, and 4.02%, respectively. In addition to A_s, large amount of reduction in PRE are noticed using proposed method which are 65.62%, 38.8%, 67.10%, 99.53%, 7.55%, and 42.7%, respectively.

5 Conclusions

In this paper, a new algorithm based on multi-swarm PSO is presented to design the linear phase prototype filter of QMF bank. To avoid the premature convergence in higher order filter design, SA is hybridized with multi-swarm PSO. The performance of the proposed SAMCPSO algorithm is compared with that of different PSO variants, ABC, BAT algorithm, and other well-known recently reported methods.

Table 1 Performance parameters obtained using proposed SAMCPSO for higher order filter banks

N	α	β	E_p	E_s	E_t	e_{am}	PRE (dB)	A_s(dB)
48	0.8	10^{-06}	3.9×10^{-09}	1.6×10^{-08}	1.4×10^{-11}	2.1×10^{-03}	0.00305	59.28
56	0.6	10^{-06}	1.6×10^{-09}	1.6×10^{-09}	4.8×10^{-12}	1.2×10^{-03}	0.00135	67.16
64	0.5	10^{-06}	1.1×10^{-09}	6.5×10^{-10}	1.9×10^{-11}	9.1×10^{-04}	0.00105	77.135
76	0.5	10^{-07}	2.7×10^{-10}	6.4×10^{-11}	7.5×10^{-12}	7.7×10^{-04}	0.00096	102.83
82	0.3	10^{-08}	1.4×10^{-11}	2.7×10^{-13}	1.1×10^{-18}	6.8×10^{-04}	0.00082	112.88
90	0.01	10^{-09}	1.4×10^{-11}	9.6×10^{-14}	7.7×10^{-19}	5.7×10^{-04}	0.00042	119.46
100	0.01	10^{-09}	2.9×10^{-12}	8.6×10^{-14}	2.7×10^{-19}	4.8×10^{-04}	0.00033	122.24

Table 2 Performance comparison of proposed SAMCPSO for higher order (N = 48 and 82) filter bank

N	Method	A_s (dB)	e_{am}	E_p	E_s	E_t	PRE (dB)
48	[4]	50.55	–	2.54×10^{-09}	6.62×10^{-08}	3.87×10^{-27}	0.0089
	[5]	51.38	–	2.70×10^{-08}	8.91×10^{-08}	1.93×10^{-22}	0.0050
	ABC [11]	55.08	–	3.16×10^{-10}	2.95×10^{-08}	–	0.0093
	BAT [18]	48.18	0.1618	$4.40 \times 10^{+02}$	$9.00 \times 10^{+02}$	6.49	0.6445
	COABC [19]	57.73	3.87×10^{-03}	5.08×10^{-10}	1.77×10^{-08}	3.17×10^{-13}	0.00331
	CPSO [20]	56.99	3.9×10^{-03}	3.58×10^{-10}	1.76×10^{-10}	2.26×10^{-13}	0.00534
	SAMCPSO	**59.28**	$\mathbf{2.08 \times 10^{-03}}$	$\mathbf{3.93 \times 10^{-09}}$	$\mathbf{1.63 \times 10^{-08}}$	$\mathbf{1.36 \times 10^{-11}}$	**0.00306**
82	[4]	96.16	–	1.31×10^{-13}	8.30×10^{-13}	2.01×10^{-20}	0.0037
	[5]	96.54	–	3.37×10^{-11}	4.18×10^{-12}	2.07×10^{-14}	0.0010
	ABC [11], BAT [18], COABC [19], CPSO [20]				No feasible solution found		
	SAMCPSO	**112.88**	$\mathbf{6.86 \times 10^{-04}}$	$\mathbf{1.37 \times 10^{-11}}$	$\mathbf{2.66 \times 10^{-13}}$	$\mathbf{1.37 \times 10^{-18}}$	**0.000829**

Based on the significantly encouraging simulation results, it can be concluded that the proposed SAMCPSO method results in a better quality prototype filter. Further, SAMCPSO algorithm exhibits a better balancing ability between exploration and exploitation, and it can be applied effectively in solving other complex real-life problems. The design of multiplier-less two-dimensional digital filters and higher order Hilbert transformer using the proposed algorithm are left as future work.

References

1. Jain V, Crochiere R (1984) Quadrature mirror filter design in time domain. IEEE Trans Acoust Speech Signal Process 32(2):353–361
2. Rafi SM, Kumar A, Singh GK (2013) An improved particle swarm optimization method for multirate filter bank design. J Franklin Inst 350(4):757–769
3. Kumar A, Rafi SM, Singh GK (2012) A hybrid method for designing linear-phase quadrature mirror filter bank. Digit Signal Process 22(3):453–462
4. Kuldeep B, Singh VK, Kumar A, Singh GK (2015) Design of two-channel filter bank using nature inspired optimization based fractional derivative constraints. ISA Trans 54:101–116
5. Baderia K, Kumar A, Singh GK (2015) Design of quadrature mirror filter bank using polyphase components based on optimal fractional derivative constraints. Int J Electron Commun (AEÜ) 69(9):1254–1264
6. Sahu OP, Soni MK, Talwar IM (2006) Marquardt optimization method to design two-channel quadrature mirror filter banks. Digital Signal Process 16(6):870–879
7. Dhabal S, Venkateswaran P (2017) An efficient gbest-guided Cuckoo Search algorithm for higher order two channel filter bank design. Swarm Evol Comput 33:68–84
8. Upender J, Gupta CP, Singh GK (2010) Design of two-channel quadrature mirror filter bank using particle swarm optimization. Signal Process 20(2):304–313
9. Misra D, Dhabal S, Chakrabarti R, Venkateswaran P (2012) Canonical signed digit representation of quadrature mirror filter using genetic algorithm. In: Proceedings of international conference on communications, devices and intelligent systems (CODIS), Kolkata, pp 65–68
10. Oliveira HA, Petraglia A, Petraglia M R (2009) Design of FIR quadrature mirror-image filter banks using fuzzy adaptive simulated annealing. In: Proceedings of 13th digital signal processing workshop and 5th IEEE signal processing education workshop, Marco Island, pp 485–489
11. Agrawal SK, Sahu OP (2015) Artificial bee colony algorithm to design two-channel quadrature mirror filter banks. Swarm Evol Comput 21:24–31
12. Kumar A, Singh GK, Anand RS (2010) An improved method for designing quadrature mirror filter banks via unconstrained optimization. J Math Model Algorithms 9(1):99–111
13. Jamili A, Mohammad AS, Moghaddam RT (2011) A hybrid algorithm based on particle swarm optimization and simulated annealing for a periodic job shop scheduling problem. Int J Adv Manuf Technol 54(1):309–322
14. Niu B, Zhu Y, He X, Wu H (2007) MCPSO: a multi-swarm cooperative particle swarm optimizer. Appl Math Comput 185(2):1050–1062
15. Zhang J, Ding X (2011) A multi-swarm self-adaptive and cooperative particle swarm optimization. Eng Appl Artif Intell 24(6):958–967
16. Xu X, Tang Y, Li J, Hua C, Guan X (2015) Dynamic multi-swarm particle swarm optimizer with cooperative learning strategy. Appl Soft Comput 29:169–183
17. Hasanzadeh M, Meybodi MR, Ebadzadeh MM (2013) Adaptive cooperative particle swarm optimizer. Appl Intell 39(2):397–420

18. Yang XS (2013) Bat algorithm and cuckoo search: a tutorial. In: Artificial intelligence, evolutionary computing and metaheuristics. Studies in computational intelligence, vol 427. Springer, Berlin, Heidelberg
19. Luo J, Wang Q, Xiao X (2013) A modified artificial bee colony algorithm based on converge-onlookers approach for global optimization. Appl Math Comput 219(20):10253–10262
20. Petrović M, Mitić M, Vuković N, Miljković Z (2016) Chaotic particle swarm optimization algorithm for flexible process planning. Int J Adv Manuf Technol 85(9–12):2535–2555

Medav Filter—Filter for Removal of Image Noise with the Combination of Median and Average Filters

Sayantan Gupta and Sukanya Roy

Abstract Noise in an image is undesirable to us as it disrupts and degrades the quality of the image. Noise removal is always a difficult task so as edge preservation when the intensity of the disrupted noise in the original image is high. In this paper, we proposed the Medav Filter which is a combination of mean and adaptive median filter that optimally adjusts the level of mask operations according to the noise density. The median filter has good noise removal qualities, but its complexity is undesirable. While the mean filter is unable to remove heavy tailored noise, we see its complexity increases in the presence of noise which is dependent upon the signal. In the Medav Filter, we proposed an algorithm to improve the peak signal–to-noise ratio (PSNR) which eventually improved the signal-to-noise ratio (SNR). We also described an efficient model for image restoration. The analysis of the algorithm and the Medav Filter shows that the complexity, as well as the performance, is improved as compared to the primitive filters.

Keywords Digital Image Processing · Noise removal · Impulse noise
Dual threshold median filter · Average filter · Medav Filter · Image restoration
Computer vision

1 Image Restoration in Digital Image Processing

1.1 Literature Survey

The primary objective of Digital Image Processing is to discard the unwanted noise in an image which is irrelevant to us, by using various complex algorithms and procedures. Finally, our main intention is to extract detailed specific information from a particular image [1, 2]. The task becomes difficult in the case of computer vision in satellite images or optical images where the extent of noise is quite high and

S. Gupta (✉) · S. Roy
Computer Science Department, University of Engineering and Management, Kolkata, India
e-mail: sayantangupta999@gmail.com

S. Bhattacharyya et al. (eds.), *Recent Trends in Signal and Image Processing*,
Advances in Intelligent Systems and Computing 727,
https://doi.org/10.1007/978-981-10-8863-6_2

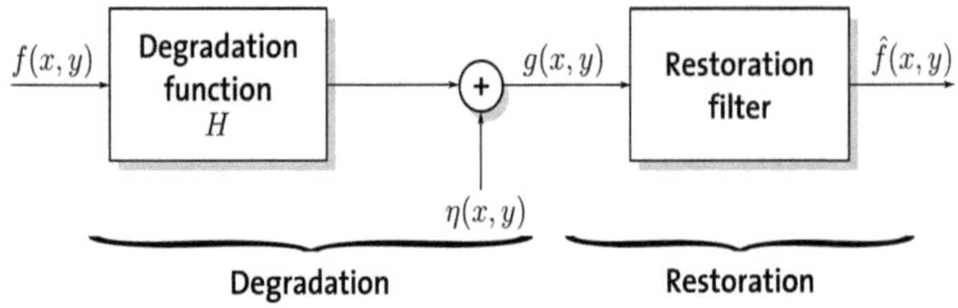

Fig. 1 Degradation and restoration model

very less amount of information can be gained from such inputs [1, 3]. The outputs which we get from the processing algorithms are reversible and clear and detailed images with less noise as compared with the input noisy image. Now, the various sources of noise can be the camera shutter, low, or less light during the capture of the image or sensitivity mode of the camera being very high, etc. [4, 5]. All these factors play an important role in the quality of the image. In a digital image, noise implies the arbitrary pixels which are scattered in the image. Noise removal basically falls in the "Pre-Processing" step in Digital Image Processing (Fig. 1).

Let us have O_i as a variable lying in the specific interval of [0, 1] which basically represents the black and white levels in an image. Let the probability of occurrence of O_i be $P_i\,(O_i)$:

$$P_i\,(O_i) = \frac{N_i}{n} \tag{1}$$

where $i = 0, 1, 2 \ldots L - 1$

Here,

L represents the number of Levels present.
N_i represents the number of times black or white appears in an image.
n represents the number of pixels present in the image.

Now, let us consider the number of bits to represent O_i be $L(O_i)$.

So, the total number of bits that will be required to represent each pixel in an image will be:

$$L_{avg} = L\,(O_i)\; * \; P_i\,(O_i) \tag{2}$$

There are various types of noise that can be embedded in an image: salt-and-pepper noise, Gaussian noise, impulse noise, shot noise, random noise, quantization noise, etc., each having its own specifications and effects in an image. However, it is not only noise which we have to consider during image processing, and motion blur and camera misfocus are also some of the reasons for the disruption of an image [5, 6].

However, all forms of noise are not always undesirable. Obviously, when the intensity of any particular noise becomes too high, it becomes undesirable to us. But, sometimes noises are intentionally induced in an image to improve the image perception—sometimes called "dither." An application of dither can be in the conversion of gray image to black and white image, where noise plays a vital role [7, 8].

1.2 *Image Restoration Using Recursive Filter*

Image restoration is the process in which a noisy image is taken as input [9], and by using various mathematical operations and functions, we reverse the process and get back the actual rectified denoised image. Image restoration is often combined with image enhancement to get a clear and enhanced image.

Model of Degradation and Restoration:
Degradation Model: Let us consider a model which is used to disrupt our image as per our need.

In Fig. 1,

(x, y) are coordinates in space (here we consider 2D),

$f(x, y)$ is the input clear image,

H is the degradation function which is responsible for the disruption in the image; it is basically a low-pass filter used for degradation, and

$n(x, y)$ is the noise added to the input image.

The output g(x, y) corresponds to:

$$g\,(x,\ y) = H\left[f\,(x,\ y)\right] + n\,(x,\ y) \tag{3}$$

Now, there are two conditions in the output function:

1. The described model is linear if this condition is satisfied:

$$H\left[k_1 f_1\,(x,\ y) + k_2 f_2\,(x,\ y)\right] = k_1 H\left[f_1\,(x,\ y)\right] + k_2 H\left[f_2\,(x,\ y)\right] \tag{4}$$

2. The model will be position or space invariant if the following is satisfied:

$$H\left[f\,(x - x_0,\ y - y_0)\right] = g\,(x - x_0,\ y - y_0) \tag{5}$$

The output disrupted image is just the twisted version of the original input image in the Degradation Model. Then, the image is transferred to the restoration filter. There are various types of such filters, but in this paper, we will describe inverse recursive filter for our purpose.

The primary concept of the recursive filter is to make a good initial guess of the input image $f(x, y)$ focusing on the output $g(x, y)$.

The basic equations are:

$$f_0\,(x,\ y) = \gamma g\,(x,\ y) \tag{6}$$

$$f_{i+1}(x, y) = f_i(x, y) + \gamma\,(g(x, y) - f_i(x, y) * H(x, y)) \quad (7)$$

where f_0 is the initial guess made by the algorithm.

γ is the factor of convergence.

Now, if the algorithm makes a good guess then f_i which is passed by $H(x, y)$ will be in close proximity with the output $g(x, y)$. Also, f_{i+1} will disappear from the equation and will converge with f_i.

If we consider the equations in the frequency state, then

$$F_0 = \gamma \acute{g}(x, y) \quad (8)$$

$$F_{i+1}(x, y) = F_i(x, y) + \gamma\left(\acute{g}(x, y) - F_i(x, y) * \acute{H}(x, y)\right) \quad (9)$$

Solving the equations recursively, we get

$$F_i(x, y) = \gamma \acute{g}(x, y)\left[1 + \left(1 - \gamma \acute{H}(x, y)\right) + \cdots + \left(1 - \gamma \acute{H}(x, y)\right)^i\right] \quad (10)$$

$$\frac{\acute{g}(x, y)}{\acute{H}(x, y)}\left[1 - \left(1 - \gamma \acute{H}(x, y)\right)^{i+1}\right] = F_i(x, y) \quad (11)$$

In the end, the $\left(1 - \gamma \acute{H}(x, y)^{k+1}\right)$ term will be zero as i will be infinity. So, we will get the desired output of the filter.

We have to choose the convergence factor (γ) in such a way that it satisfies the following condition:

$$\left|1 - \gamma \acute{H}(x, y)\right| < 1 \quad (12)$$

So, γ will always be a number lying between 0 and 1. The larger the value, the faster F_i and F_{i+1} will converge. However, choosing very large factor will lead them to get diverge which is not at all desirable.

2 Medav Filter Model Proposal—Combination of Mean and Median Filter

2.1 *Literature Review of Adaptive Median*

The adaptive median filter is the improved version of the median filter which adjusts [10] the value of the masking matrix automatically with reference to the noise levels in the matrix. In the traditional median filter, the smaller the size of the masking matrix is, the more image specifications are restored, but the performance in noise removal is drastically affected. While on the other hand if the size of the masking matrix is large, then the performance will improve but fewer image specifications will be stored [5, 7, 8].

In the adaptive median filter, let M_{max} and M_{min} be the maximum and a minimum number of gray level in an image. Let M_{avg} and M_{med} be the average and median values of the gray levels.

Now, $f(x, y)$ be the mean value of the mask and "n" be the size of the mask.

ALGORITHM:

Step 1. To adjust the mask adaptively.

- Let us consider 'n' to be 3
- $W_1 = M_{med} - M_{min}$ and $W_2 = M_{med} - M_{max}$ (Computation)
- If W_1>0 or W_2<0, then Jump to Step 2. Else enlarge the size of the masking matrix and initialize $n= n + 2$ and Return to "Computation"

Step 2. Procedure for Median Filtering.

2.2 *Proposed Medav Filter Algorithm*

The performance of noise removal in median filters is better than the mean filters in case of random noise or salt-and-pepper noise, but in our experiment, we will use impulse noise as the width of the narrow pulses is almost less than half of the size of the mask or the masking matrix.

In a digital image, there is a specific relation between current and the next pixel. Now, let us consider a pixel whose value is more than its neighboring pixels—then the considered pixel is affected by noise. In our algorithm, we check every pixel individually with its neighboring values, and if it is affected by noise, the algorithm replaces the pixel value by the median value. This algorithm not only reduces the complexity of the procedure but also retains many details of an image. The PSNR and SNR values are also improved in the case of Medav Filters.

The algorithm for the Medav Filter is:

ALGORITHM:

Step 1. Take the values of a particular pixel as input.

Step 2. Calculate the Average value of the mask input and store it in V_{AVG}.

Step 3. If the value of the input pixel is greater than V_{AVG}, then find the Median value and replace the value of the pixel with it, otherwise retain the original value of the pixel.

Step 4. Repeat until all pixels are checked.

Analysis of the Algorithm

Let the pixel value be (x, y), $f(x, y)$ be the input image and the masking matrix be of size 3, with the average value being $\acute{f}(x, y)$.

Now, calculating by the primitive Median algorithm:

$$V_{avg} = \frac{[f(x-1, y) + \cdots f(x, y) + f(x, y+1) + \cdots f(x+1, y+2)]}{9} \tag{13}$$

Now, the median value should be,

$$f(x-1, y) + \cdots f(x, y) + f(x, y+1) + \cdots f(x+1, y+2) \tag{14}$$

Now, in our algorithm, the value of $f(x, y)$ is replaced by $\acute{f}(x, y)$. So,

$$V_{AVG} = \left[\frac{f(x-1, y) + \cdots \acute{f}(x, y) + f\left(x, y+\acute{1}\right) \cdots f(x+1, y+2)}{9}\right] \tag{15}$$

So, we see that $V_{AVG} < V_{avg}$, so we can effectively preserve certain specification of the image and the time complexity of the algorithm being $O(n)$ is greatly improved.

2.3 Performance Evaluation of the Medav Filter

The noise reduction evaluation of the filter can be done in various ways, but in our paper, we choose peak signal-to-noise ratio (PSNR) and signal-to-noise ratio (SNR) as evaluation parameters, which are universally accepted [11].

Let the input image be $f(x, y)$ and the size of the image being $I \times J$, while the output image from the Medav Filter be $O(x, y)$ and size $I \times J$. Then, we have

$$MSE = \left[\frac{1}{I \times J}\sum_{i=1}^{I}\sum_{j=1}^{J}\{O(x, y) - f(x, y)\}^2\right] \tag{16}$$

The peak signal-to-noise ratio will be

$$PSNR = \frac{10 \log \delta_{max}^2}{MSE} \; dB \tag{17}$$

Here, $\delta_{max} = 2^i - 1$, where i is the total number of pixels present in binary form:

$$Val = \left\{\sum_{i}^{I}\sum_{j}^{J} f(x, y)^2\right\} / \sum_{i}^{I}\sum_{j}^{J}\{O(x, y) - f(x, y)\}^2 \tag{18}$$

Table 1 Experimental comparison of various filters with Medav Filter

Noise density	Primitive median filter		Adaptive median filter		Medav Filter model	
Levels (%)	PSNR(dB)	SNR(dB)	PSNR(dB)	SNR(dB)	PSNR(dB)	SNR(dB)
10	31.0289	25.3748	31.1573	26.1009	32.1573	26.1947
35	27.1486	21.5319	27.9651	21.8701	31.3329	25.5469
50	19.4371	12.0616	19.7662	12.6719	30.6591	24.9017
60	–	–	–	–	29.8258	21.0001
80	–.	–	–	–	27.1295	20.7861

$$SNR = 10 \log Val \quad dB \tag{19}$$

The experimental calculations and comparison of traditional median, adaptive median filter, and the Medav Filter are shown in Table 1.

2.4 *Simulation of the Medav Filter*

We simulated the Medav Filter using MATLAB 2007b by inducing kinds of impulse noises. The results are implemented with the inbuilt "Lena" Image.

The simulation results of the median, adaptive median, and Medav Filter are shown in Fig. 2.

3 Conclusion and Discussions

As we see the Medav Filter is a combination of mean and median filters, it has a huge advantage over the traditional Median Filters. The edge preservation ratio for the Medav Filters is also very high. We can also preserve the details of the image more specifically by using such filtering methods. The time complexity of the algorithm has also been improved and stimulated. The edge preservation quantifiers are of great use in such image restoration because such hybrid filters can effectively suppress noise and increase the efficiency to a great extent as compared with the traditional algorithms. The future of Digital Image Processing lies in the hands of developing efficient algorithms which can confess as much information from an image. Recent developments in Quantum Digital Image Processing are some of the innovative steps taken in this vast and popular field.

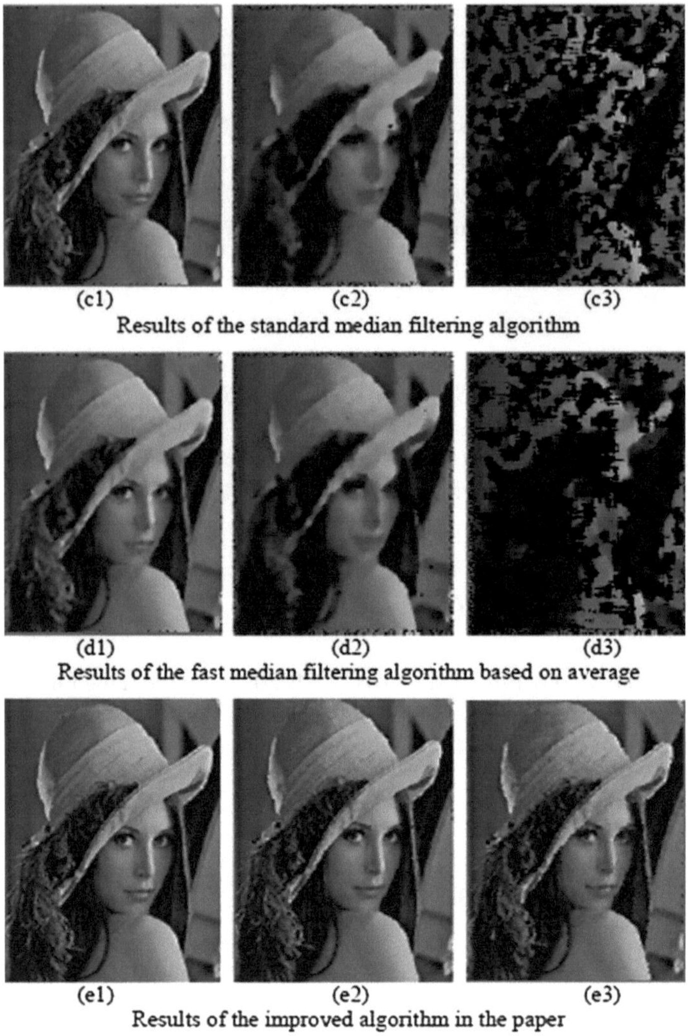

Fig. 2 Simulation results of filter comparison

References

1. Liu G, Guo W (2010) Application of improved arithmetic of median filtering denoising. Comput Eng Appl 46(10):187–189
2. Wang X, Li F (2010) Improved adaptive median filtering. Comput Eng Appl 46(3):175–176
3. Huang Q, Zhou H, Feng H (2002) A fast and effective algorithm of pulse noise filtering for imaging data. Comput Eng Appl (17)
4. Huang TS, Tang GT (1979) A fast two-dimensional median filtering algorithm. IEEE Trans Acous Speech Signal Process 27(1):13–18

5. Gupta S, Sau K, Bhattacharya S, Chatterjee T (2017) Quantum brain-time matrix—the correlation between brain-mind-time with quantum spin and quantum entanglement: quantum computation of the brain and representation as a matrix. In: 2017 8th annual industrial automation and electromechanical engineering conference (IEMECON), Bangkok, Thailand, pp 168–174. https://doi.org/10.1109/iemecon.2017.8079583
6. Chubin W (2006) The fast algorithm of medium filtering based on mean. J Jiangsu Teachers Univ Technol 12(6):102–106
7. Wang C, Ye Z (2008) Salt-and-pepper noise removal by adaptive median filter and TV inpainting. J Univ Sci Technol China 38(3)
8. Chen T, Xia L (1994) Digital image processing. Posts & Telecommunications Press, Beijing
9. Gupta S, Sau K, Pramanick J, Pyne S, Ahamed R, Biswas R (2017) Quantum computation of perfect time-eavesdropping in position-based quantum cryptography: quantum computing and eavesdropping over perfect key distribution. In: 2017 8th annual industrial automation and electromechanical engineering conference (IEMECON), Bangkok, Thailand, pp 162–167. https://doi.org/10.1109/iemecon.2017.8079582
10. Gupta S, Mohanta S, Chakraborty M, Ghosh S (2017) Quantum machine learning-using quantum computation in artificial intelligence and deep neural networks: quantum computation and machine learning in artificial intelligence. In: 2017 8th annual industrial automation and electromechanical engineering conference (IEMECON), Bangkok, Thailand, pp 268–274. https://doi.org/10.1109/iemecon.2017.8079602
11. Kundu S, et al (2016) Quantum computation: from Church-Turing thesis to qubits. In: 2016 IEEE 7th annual ubiquitous computing, electronics & mobile communication conference (UEMCON), New York, NY, pp 1–5. https://doi.org/10.1109/uemcon.2016.7777805

Classification of Metamaterial-Based Defected Photonic Crystal Structure from Band-Pass Filter Characteristics Using Soft Computing Techniques

Soumen Mukherjee, Arup Kumar Bhattacharjee, Payel Halder and Arpan Deyasi

Abstract The present paper deals with the classification problem of metamaterial-based photonic crystal from its band-pass filter characteristics obtained experimentally in presence and absence of defects at optical communication spectrum of 1.55 μm. Two well-known DNG materials namely paired nanorod (n = −0.3) and nano-fishnet with elliptical void (n = −4) are considered for analysis purpose, and presence of point defects is taken into account in otherwise ideal structure which makes it as a four-class problem. Band-pass filter characteristics are measured for all the classes for both normal and oblique incidences separately with dimensional and incident angle variations; different soft computing techniques are applied for classification purpose as it is hardly possible to identify the device from the filter behavior. Apriori algorithm is utilized for association analysis to determine 100% confidence. Result shows that 98.53% accuracy is provided with neural network-based classifier with 98.93% sensitivity and 98.08% specificity when computation is made over 1000 samples.

Keywords Transmittivity · Photonic crystal · Defect · Neural network
Sensitivity · Specificity · Metamaterial

S. Mukherjee · A. K. Bhattacharjee
Department of Computer Application, RCC Institute of Information Technology, Kolkata, India
e-mail: soumou601@gmail.com

A. K. Bhattacharjee
e-mail: arupk.b@gmail.com

P. Halder · A. Deyasi (✉)
Department of Electronics and Communication Engineering, RCC Institute of Information Technology, Kolkata, India
e-mail: deyasi_arpan@yahoo.co.in

P. Halder
e-mail: payelh139@gmail.com

S. Bhattacharyya et al. (eds.), *Recent Trends in Signal and Image Processing*,
Advances in Intelligent Systems and Computing 727,
https://doi.org/10.1007/978-981-10-8863-6_3

1 Introduction

One-dimensional photonic crystal [PhC] is the subject of research in the last 30 years following the pioneering work of Yablonovitch [1] after the path-breaking theoretical research of Loudon [2] regarding the propagation of electromagnetic wave in periodic dielectric medium. Owing to the formation of electromagnetic bandgap [3] inside the structure due to periodic variation of refractive indices of the constituent materials, the structure is already utilized in making of photonic transmitter [4], receiver [5], sensor [6], switch [7], fiber [8], quantum information processing [9], etc. The role of materials and their structural parameters along with mode of propagating wave play key aspects in shaping the characteristic of PhC-based devices [10, 11].

Metamaterial-based PhC is the subject of research in the last few years due to its unique feature of guiding electromagnetic wave inside [12], where theoretical foundation is based on Maxwell–Garnett effective medium theory. Mutual coupling effect is observed for Si nanopillars-based plasmonic waveguide [13]. Using metamaterial, antennas are already designed where photonic crystal property is exhibited [14]. This clearly speaks in favor of DNG material-based PhC design than conventional Si–SiO_2 counterpart [15].

In the present paper, classification is made between two different types of DNG material-based PhC structure from the transmittivity analysis as both the type of structures is applicable for optical filter design. Paired nanorod and nano-fishnet structure with elliptical void are considered for analysis purpose, and filter characteristics are measured for different structural parameters and incident angles under TE and TM mode propagations separately. Different soft computing techniques are applied for the classification of the structures and also for distinguishing between defected with corresponding ideal structures by only the data generated by the photonic crystals, which can be useful in cases where the identification of the composition of photonic crystal is not possible outside of the component structure. Association analysis is also performed for confidence rule evaluation between different pair of attributes. Accuracy of the obtained result speaks in favor of the classification which is otherwise impossible from the experimental result.

2 Mathematical Modeling

Considering the phase factor of the field propagating through uniform medium, propagation matrix is given as the function of barrier and well widths

$$P_{1,2} = \begin{pmatrix} \exp[jk_{1,2}d_{1,2}] & 0 \\ 0 & -\exp[jk_{1,2}d_{1,2}] \end{pmatrix} \tag{1}$$

where $d_{1,2}$ is the dimension of barrier/well layer and $k_{1,2}$ is the propagation vector. Considering 'f' as normalized defect density, propagation matrix in presence of defect is given by

$$P_{1,2} = \begin{pmatrix} \exp[jk_{1,2}d_{1,2}]f & 0 \\ 0 & -\exp[jk_{1,2}d_{1,2}]f \end{pmatrix} \quad (2)$$

Thus, transfer matrix for the elementary cell (constituting of one barrier and one well layer) is

$$M = M_1^T P_1 M_2^T P_2 \quad (3)$$

Transfer matrix for the elementary cell in presence of defect in second layer (well) is

$$M_{defect} = M_1^T P_1 M_2^T P_{2defect} \quad (4)$$

where M is the transfer matrix between the adjacent layers, given by

$$M_{1,2}^T = \frac{1}{t}\begin{pmatrix} 1 & r_{21,12} \\ r_{21,12} & 1 \end{pmatrix} \quad (5)$$

For a perfectly periodic medium composed of N such elementary cells, total transfer matrix for such a structure is

$$M_{tot} = M_N \quad (6)$$

For a defected periodic medium composed of N such elementary cells, the total transfer matrix for such a structure is

$$M_{tot} = M_{defect_N} \quad (7)$$

Transmission coefficient is given by

$$T = \frac{1}{M_{11}^2(tot)} \quad (8)$$

In absence of defect, transmission coefficient will be calculated from $P_{1,2}$.

Table 1 Classification result of metamaterial with four classes and three features

Classifier type	Accuracy (%)	Time of training (s)
Neural network (proposed work)	55.27	1.13421
Complex tree	58.2	0.44728
Weighted KNN	60.5	0.28869

Table 2 Classification result of metamaterial with two classes and 15 features

Classifier type	Accuracy (%)	Time of training (s)
Neural network (proposed work)	98.53	0.95462
Complex tree	98.0	0.36141
Linear SVM	91.2	0.55351

3 Results and Discussions

Based on the analysis of Sect. 2, transmittivity profile can be obtained for different types of structures in presence or absence of defect. Now as per the filter characteristics obtained, the data set can be classified as of two classes of metamaterial namely paired nanorod (M1) and nano-fishnet with elliptical void (M2) comprising with another two classes, in presence and absence of point defect in both the structures, making it altogether four-class problem. In each class, 1000 samples are considered from experimental results. As per practical aspect, it is rational to distinguish between defected structures, based on their characteristic obtained under different mode of propagation, and also for different structural parameters. Since it is never possible to differentiate between the two metamaterials from the filter profile, henceforth this classification problem is solved considering 12 features with 1000 samples in each class. The features for each type of structure are based on different layer widths [lower, medium, and higher] and also of incident angles for TM and TE mode propagations, respectively. When defected structure is compared with the ideal one in terms of performance, then the two-class classification problem becomes four-lass problem with 4000 samples are present, 1000 samples in each four-class. Initially, the classification learner toolbox of MATLAB and neural network toolbox are used to classify the two-class and four-class problem with different classifiers. The detailed results are given in Table 1 and Table 2, respectively.

It can be seen from the table that two-class classification yields better result with highest accuracy of 98.53% using neural network classifier. In two-class classification problem, complex tree and linear support vector machine give an accuracy of 98% and 91.2%, respectively. In four-class classification as the number of feature is very less (three features), it yields not satisfactory result, with only 60.5% accuracy using weighted KNN classifier. Other two classifiers, i.e., neural network and complex tree, give 55.27 and 58.2% accuracy. All the above classifications are done with fivefold

Table 3 Performance metric of neural network classifier

Accuracy	Sensitivity	Specificity	False positive rate	False negative rate	Precision
98.53%	98.93%	98.08%	1.072%	1.918%	98.22%

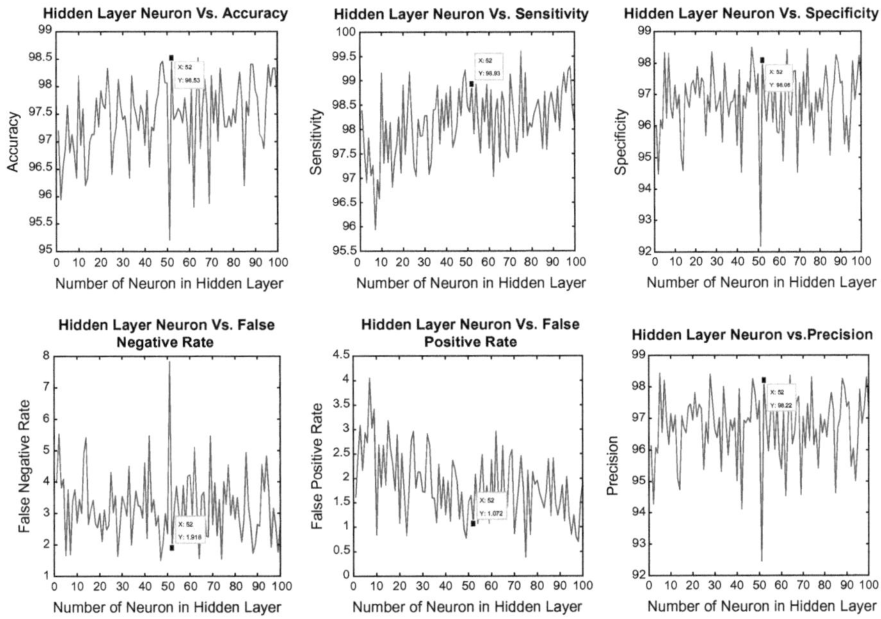

Fig. 1 Hidden layer neuron versus **a** accuracy; **b** sensitivity; **c** specificity; **d** false positive rate; **e** false negative rate; **f** precision

cross-validation. In neural network, 'Scaled Conjugate Gradient Back-propagation' training function and mean squared error function with single layer are used.

The two-class classification problem is discussed here in detail. The multilayer neural network (MLP) is used to classify the two-class classification problem with 2000 samples (1000 sample from each class) and 15 features. The present work is done with varying number of neuron from 1 to 100 with a single hidden layer to find the best accuracy result. In the present work, total 70% (1400) samples are taken for training, 15% (300) samples are taken for validation, and 15% (300) samples are for testing. The best accuracy of 98.53% is found with hidden neuron size 52. The details of performance metric of the neural network classifier are given in Table 3.

In Fig. 1, variations of different performance metric in percentage with number of hidden layer neuron are shown. It can be noted that the variation of result with number of hidden layer neuron is very less (4–5%).

In Fig. 2 and Fig. 3, respectively, the confusion matrix and the receiver operating characteristic (ROC) curve are shown for a single instance of the neural network classifier. The confusion matrix shows whether the classification by the classifier

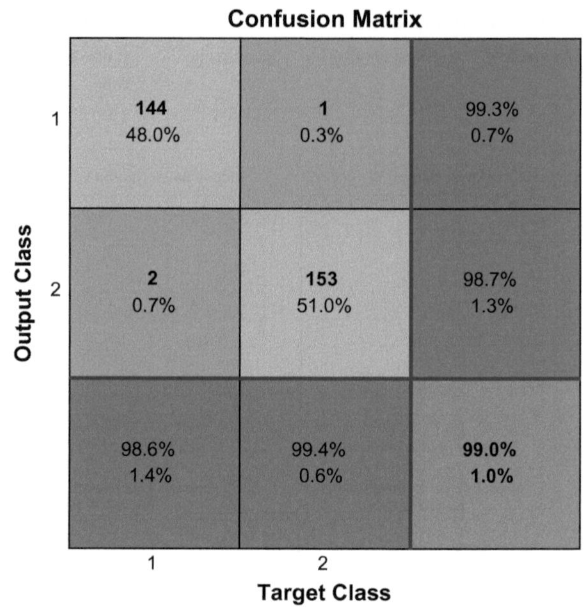

Fig. 2 Confusion matrix of two-class classification problem

falls in the proper class or there is a misclassification. In this two-class classification problem, the confusion matrix is a 2 × 2 matrix, where the (1, 1) and (2, 2) cell of the confusion matrix shows the percentage of accurate classification and the (1, 2) and (2, 1) cell shows the percentage of misclassification. The ROC curve represents a curve between the true positive rate and the false positive rate of each class.

It can be seen that the area under the curve is nearly 1 in both the class, which means the result found is of high accuracy.

Figure 4 shows that the neural network achieves best validation performance of 0.035295 at epoch 42 with cross-entropy error function. With six transformed feature generated by principal component analysis (PCA), the present system achieves an accuracy of 98.73%. Figure 5 shows the accuracy versus the PCA component curve of the two-class classification problem.

Knowledge Discovery in Database (KDD) also known as data mining is used to determine patterns and trends to predict outcomes [16]. In this section, we have tried to apply association rules of data mining to the data available regarding metamaterials. Association rules are the rule-based machine learning techniques used for mining frequent item sets and generate relevant association rules [17]. The underlying principle of association rules is to operate on a database usually containing information about transactions, for instance items purchased by customers from a store (Market Basket Analysis).

In the present problem, after preprocessing, Apriori algorithm [18] is applied to this binary data set to generate the rules. Since number of features has 15 columns and class has 1 column, so binary representation of features is $15 \times 2 = 30$; i.e., Attribute

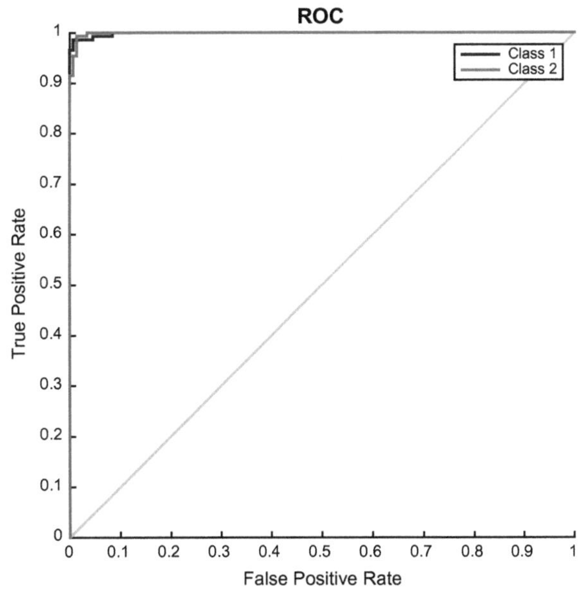

Fig. 3 ROC curve of two-class classification problem

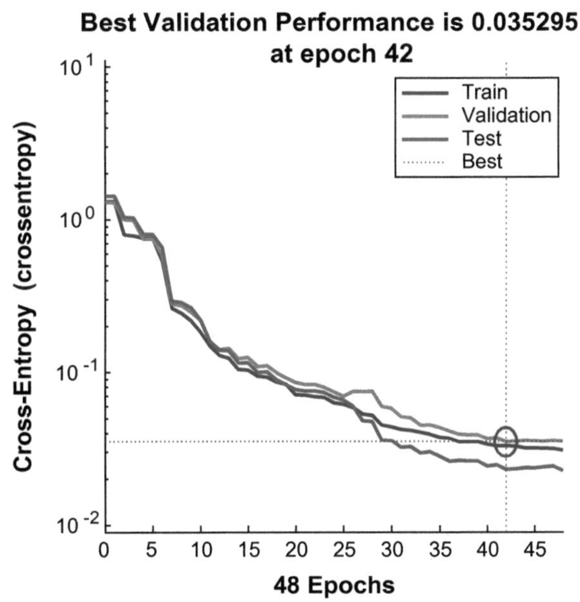

Fig. 4 Validation performance curve

no. 1 is represented as Attribute no. 1 and Attribute no. 2; similarly Attribute no. 15 is represented as Attribute no. 29 and Attribute no. 30, and class column is represented in column 31 and 32. In this column 32, data apriori algorithm is applied with the following parameter values:

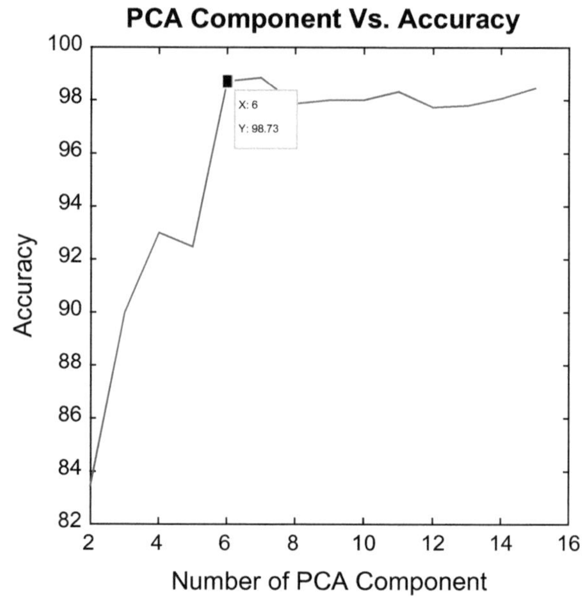

Fig. 5 PCA component versus accuracy of two-class classification

[i] Minimum Support = 0.1
[ii] Minimum Confidence = 0.7
[iii] Number of rules to generate = 1000

The concepts of closed frequent item sets are used to remove redundant rules. And the top 15 rules with class as consequence based on descending order of support with 100% confidence [19] are:

$$1,\ 29 \rightarrow 31;\ 1,\ 3 \rightarrow 31;\ 1,\ 5 \rightarrow 31;\ 1,\ 17 \rightarrow 31;\ 1,\ 7 \rightarrow 31;$$
$$2,\ 6 \rightarrow 32;\ 2,\ 4 \rightarrow 32;\ 1,\ 20 \rightarrow 31;\ 2,\ 15 \rightarrow 32;\ 1,\ 15 \rightarrow 31;$$
$$1,\ 21 \rightarrow 31;\ 2,\ 8 \rightarrow 32;\ 1,\ 14 \rightarrow 31;\ 1,\ 24 \rightarrow 31;\ 1,\ 26 \rightarrow 31;$$

Significance of the rule 1, 29 → 31 is that Attribute 1 (1/2 = 1) and Attribute 14 (29/2 = 14) are determining the class (31/2 = 16) with highest support and 100% confidence among all rule with class as consequence. Similarly rule 2 states, 1, 3 → 31; i.e., the Attribute 1 (1/2 = 1) and Attribute 3 (3/2 = 2) are determining the class (31/2 = 16) with second highest support and 100% confidence among all rule with class as consequence and so on. Apriori algorithm has generated many rules among which only those are considered whose consequence is class, i.e., column 31 and column 32.

4 Conclusion

The photonic crystal classification problem has the critical importance in optical communication engineering as the all-optical filter based on metamaterial can effectively control the SNR of the system, and a slight variation in the passband or ripple in passband will greatly affect the system performance. Since the characteristic is governed by the layer dimensions of 1D PhC and also the incident angle, hence the classification problem is based on those features for the defected structure, and distinction is also made with the ideal one. Accuracy (98.53%) is obtained for the former problem using neural network, whereas for the latter case, weighted KNN provides 60.2% only. Therefore, performance metric elements for the defected structure are plotted with hidden layer neuron, and PCA is applied for the second case which yields 98.73% accuracy when calculation is performed with best six transformed features. Association analysis is executed for 100% confidence performance, where attributes 1 and 14 are the two critical parameters in this regard. Results are extremely useful in high-frequency communication system design.

References

1. Yablonovitch E (1987) Inhibited spontaneous emission in solid-state physics and electronics. Phys Rev Lett 58:2059–2061
2. Loudon R (1970) The propagation of electromagnetic energy through an absorbing dielectric. J Phys A 3:233–245
3. Andreani LC, Agio M, Bajoni D, Belotti M, Galli M, Guizzetti G, Malvezzi AM, Marabelli F, Patrini M, Vecchi G (2003) Optical properties and photonic mode dispersion in two-dimensional and waveguide-embedded photonic crystals. Synth Met 139:695–700
4. D'Orazio A, De Palo V, De Sario M, Petruzzelli V, Prudenzano F (2003) Finite difference time domain modeling of light amplification in active photonic bandgap structures. Prog Electromagn Res 39:299–339
5. Kalchmair S, Detz H, Cole GD, Andrews AM, Klang P, Nobile M, Gansch R, Ostermaier C, Schrenk W, Strasser G (2011) Photonic crystal slab quantum well infrared photodetector. Appl Phys Lett 98:011105
6. Belhadj W, AbdelMalek F, Bouchriha H (2006) Characterization and study of photonic crystal fibres with bends. Mater Sci Eng C 26:578–579
7. Mahmoud MY, Bassou G, Taalbi A, Chekroun ZM (2012) Optical channel drop filters based on photonic crystal ring resonators. Opt Commun 285:368–372
8. Limpert J, Liem A, Reich M, Schreiber T, Nolte S, Zellmer H, Tünnermann A, Broeng J, Petersson A, Jakobsen C (2004) Low-nonlinearity single-transverse-mode ytterbium-doped photonic crystal fiber amplifier. Opt Express 12:1313–1319
9. Azuma H (2008) Quantum computation with kerr-nonlinear photonic crystals. J Phys D Appl Phys 41:025102
10. Golmohammadi S, Moravvej-Farshi MK, Rostami A, Zarifkar A (2007) Spectral analysis of the fbonacci-class one-dimensional quasi-periodic structures. Prog Electromagnet Res 75:69–84
11. Rojas JAM, Alpuente J, L´opez-Esp´i P, Garc´ia P (2007) Accurate model of electromagnetic wave propagation in unidimensional photonic crystals with defects. J Electromagnet Waves Appl 21:1037–1051
12. Vasic B, Isic G, Gaji R, Hinger K (2010) Controlling electromagnetic fields with graded photonic crystals in metamaterial regime. Opt Express 18:20321–20333

13. Butt H, Dai Q, Wilkinson TD, Amaratunga GAJ (2011) Photonic crystals & metamaterial filters based on 2D arrays of silicon nanopillars. Prog Electromagnet Res 113:179–194
14. Dorrani Z, Mansouri-Birjandi MA (2013) Transmitted antenna in photonic crystals with negative refraction. Int Res J Appl Basic Sci 5:1431–1437
15. Deyasi A, Majumder A (2016) Design of DNG material-based photonic bandpass filter at 1.55 μm with oblique incidences of electromagnetic wave. In: IEEE international conference on microelectronics, computing and communication, pp 1–5
16. Han J, Kamber M (2006) Data mining: concepts and techniques, 2nd edn. China Machine Press, 4
17. Agrawal R, Imieliński T, Swami A (1993) Mining association rules between sets of items in large databases. In: Proceedings of the 1993 ACM SIGMOD international conference on Management of data, 207
18. Yabing J (2013) Research of an improved apriori algorithm in data mining association rules. Int J Comput Commun Eng 2(1)
19. Li L, Shaorong F, Yongsheng X (2007) Mining frequent item sets from several conditional FP -trees. J Comput Eng Appl 43(5):175–177

Sparse Encoding Algorithm for Real-Time ECG Compression

Rohan Basu Roy, Arani Roy, Amitava Mukherjee, Alekhya Ghosh, Soham Bhattacharyya and Mrinal K. Naskar

Abstract In this paper, we propose a sparse encoding algorithm consisting of two schemes namely geometry-based method (GBM) and the wavelet transform-based iterative thresholding (WTIT). The sub-algorithm GBM reduces the minimal ECG voltage values to zero level. Subsequently, WTIT encodes the ECG signal in time-frequency domain, obtaining high sparsity levels. Compressed Row Huffman Coding (CRHC) algorithm converts the sparse matrices into compressed, transmittable matrices. The performance of the algorithms is validated in terms of compression ratio (CR), percentage RMS difference (PRD), and time complexity.

Keywords Sparse matrix · Real-time ECG compression · Wavelet transform Iterative thresholding · Transmittable matrix

---

R. B. Roy (✉) · A. Ghosh · S. Bhattacharyya
Institute of Radio Physics and Electronics, University of Calcutta, Kolkata, India
e-mail: rohanbasuroy@gmail.com

A. Ghosh
e-mail: alex.burdwan@gmail.com

S. Bhattacharyya
e-mail: soham.bhattacharyya007@gmail.com

A. Roy · M. K. Naskar
Department of ETCE, Jadavpur University, Kolkata, India
e-mail: araniroy.ju@gmail.com

M. K. Naskar
e-mail: mrinaletce@gmail.com

A. Mukherjee
IBM India Private Limited, Kolkata, India
e-mail: amitava.mukherjee@in.ibm.com

S. Bhattacharyya et al. (eds.), *Recent Trends in Signal and Image Processing*,
Advances in Intelligent Systems and Computing 727,
https://doi.org/10.1007/978-981-10-8863-6_4

1 Introduction

A typical ECG monitoring device generates volumes of digital data creating a necessity for efficient compression before real-time transmission. Most of the prior research [1, 2] assumed ECG signal to be noiseless and would possess definite sparsity levels. However, during ECG signal acquisition, the sparsity level may vary due to mass motion, powerline interferences, and baseline drifts. The sparsity of ECG signal is examined in our algorithm involving two schemes: GBM and WTIT, followed by a lossless encoding scheme, CRHC. Here, we have achieved high CR and low PRD simultaneously, unlike [1–3]. Also, previous work [4] on ECG compression using wavelet transform assumed an arbitrary threshold value which may change due to various cardiovascular diseases. Here, we develop an end-to-end model, which (i) filters out noise from the raw ECG signal, (ii) encodes the signal in form of highly sparse matrices, (iii) converts the sparse matrices into transmittable matrices of much reduced size, (iv) reconstructs the signal using inverse transformations.

For real-time signal transmission, the processing time plays the key role. The time complexity of our proposed algorithms is in the order of $O(n)$, while most the previous related works had time complexities in the order of $O(n^2)$ (Figs. 1 and 2).

The continuous wavelet transform $\boldsymbol{X}(a, b)$ of any signal $\boldsymbol{x}(t)$ at time t can be represented as:

$$\boldsymbol{X}\,(a, b) = \frac{1}{\sqrt{a}} \int_{-\infty}^{\infty} \boldsymbol{x}\,(t)\,\boldsymbol{\Psi}^* \frac{(t-b)}{a} dt, \quad -\infty < t < \infty \tag{1}$$

Here, a is the scaling factor, b is the translation factor, $\boldsymbol{\Psi}^*(t)$ is the complex conjugate of the mother wavelet $\boldsymbol{\Psi}(t)$.

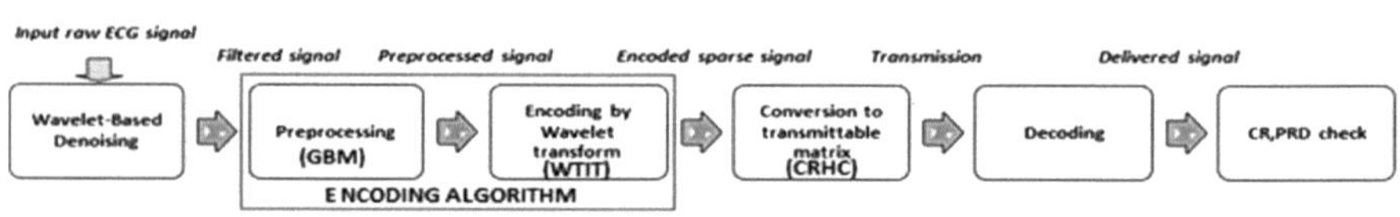

Fig. 1 Block diagram of the whole process

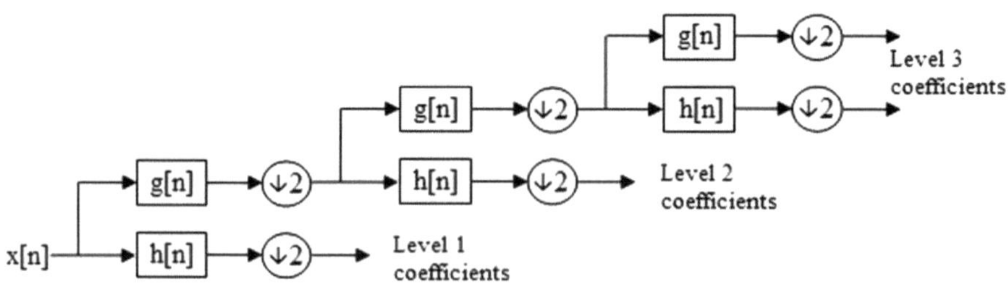

Fig. 2 DWT of x[n] up to three levels of decomposition

In discrete wavelet transform, the signal $\boldsymbol{x}[n]$, which is a discrete time signal such that $\boldsymbol{x}[n] = \boldsymbol{x}(nT)$ where $\frac{1}{T}$ is the sampling frequency, is continuously passed through a set of high-pass and low-pass filters, $\boldsymbol{h}[n]$ and $\boldsymbol{g}[n]$, respectively.

The output of the high-pass filter after down sampling by a factor of two yields detailed coefficients ($\boldsymbol{D}$) and the output of the low-pass filter after down sampling by a factor of two yields approximate coefficients ($\boldsymbol{A}$).

$$\boldsymbol{D}_j = \sum_{k=-\infty}^{\infty} \boldsymbol{h}[k]\, \boldsymbol{A}_{j-1}[2n-k] \tag{2}$$

$$\boldsymbol{A}_j = \sum_{k=-\infty}^{\infty} \boldsymbol{g}[k]\, \boldsymbol{A}_{j-1}[2n-k] \tag{3}$$

where $0 \leq n \leq l$, $\boldsymbol{D}_j$, and $\boldsymbol{A}_j$ represents jth level detailed and approximate coefficients, respectively, $0 \leq j \leq log_2 l$. l is the length of $\boldsymbol{x}[n]$.

1.1 Denoising

Wavelet-based denoising is done on vectors $\boldsymbol{A}_j$ and $\boldsymbol{D}_j$ ∀j. Decomposition-level-dependent Donoho's universal thresholding is applied:

$$T_l = \sigma\sqrt{2\log(l)} \tag{4}$$

where σ is noise variance of the signal at each level of decomposing.

Let for a given signal $\boldsymbol{u}[n]$ be an orthogonal matrix $\boldsymbol{W}$, such that the discrete wavelet-transformed matrix $\boldsymbol{U}[n]$ is $\boldsymbol{U} = \boldsymbol{W}\boldsymbol{u}$. If the additive noise to signal $\boldsymbol{x}[n]$ be $\boldsymbol{y}[n]$ which has a noise variance σ^2. Hence, the resultant signal $\boldsymbol{g}[n]$ is now:

$$\boldsymbol{g}[n] = \boldsymbol{x}[n] + \boldsymbol{y}[n] \tag{5}$$

The inverse wavelet transform can be easily computed as, $\boldsymbol{u} = \boldsymbol{W}^T\boldsymbol{U}$ The wavelet-transformed matrix $\boldsymbol{U}[n]$ is now modified to $\tilde{\boldsymbol{U}}[n]$ by applying threshold, and as a result, $\boldsymbol{u}[n]$ becomes $\tilde{\boldsymbol{u}} = \boldsymbol{W}^T\tilde{\boldsymbol{U}}$. Reduction of wavelet-transformed coefficients produces significantly noise-free estimates, which can be done either by hard or soft thresholding.

$$T_{hard}(\tilde{\boldsymbol{U}};\, T_l) = \boldsymbol{U}\,O(|\boldsymbol{U}| > T_l) \tag{6}$$

$$T_{soft}(\tilde{\boldsymbol{U}};\, T_l) = sgn(\boldsymbol{U})(\boldsymbol{U} - T_l)O(|\boldsymbol{U}| > T_l) \tag{7}$$

O is the usual indicator function and sgn represents signum function. For noise removal, we consider soft thresholding of wavelet coefficients as hard thresholding can result into significant information loss from the original signal. The noise-free ECG signal is free from zero mean band-limited Gaussian noise.

1.2 Encoding Algorithm

Preprocessing using GBM: From physiological point of view, R-R interval of an ECG signal is less than 1.2 s on the time axis and only magnitude of R wave is greater than 0.5 mV. This observation was used to build GBM.

Let us consider a sensing vector $\boldsymbol{V}$ (Fig. 3b) from the peak value of one R waveform to the peak of the next that rotates in a clockwise direction to trace out the data points on the ECG signal by varying its radial distance ($\boldsymbol{r}$). $\boldsymbol{V}$ senses P, Q, R, S, T waveforms based on their respective threshold values. When $\boldsymbol{V}$ fails to sense them, it sets the data points to zero. The clinically important features of the waveforms remain intact through GBM. As soon as $\boldsymbol{V}$ senses a T waveform, it starts rotating from the adjacent R waveform's peak and this process continues. GBM can be designed such that it senses only one among a certain number of values periodically to reduce the computation time.

Let p be the location at which peak value of ECG signal q occurs. Then, the angular displacement $\boldsymbol{\theta}$ of $\boldsymbol{V}$ can be measured by the following expression:

$$tan\boldsymbol{\theta} = [\boldsymbol{x}\,[\mathrm{n}] - q]\,/\,(\mathrm{n} - p) \tag{8}$$

$\boldsymbol{x}[\mathrm{n}]$ changes with n but p, q remains constant if $\boldsymbol{\theta} < \frac{\pi}{2}$. Else, p, q becomes abscissa and ordinate of the peak value of next R waveform, respectively. Thus, n is a variable in temporal domain and p, q are variables in spatial domain.

Lemma *Thresholding by GBM is faster to compute than a linear way in an ascending order of sample number along the total length of the ECG signal.*

Proof For a given rotation $\boldsymbol{\theta}$ of $\boldsymbol{V}$, let the linear distance covered by $\boldsymbol{V}$ be $\boldsymbol{s}$.

$\boldsymbol{r} = f(\boldsymbol{x}[n])$. Now, $\boldsymbol{r}\,(t)^2 = (q - \boldsymbol{x}\,(t))^2 + (p - t)^2$. In discrete time domain, $\boldsymbol{r}[n]^2 = (q - \boldsymbol{x}[n])^2 + (p - n)^2$. Differentiating r,

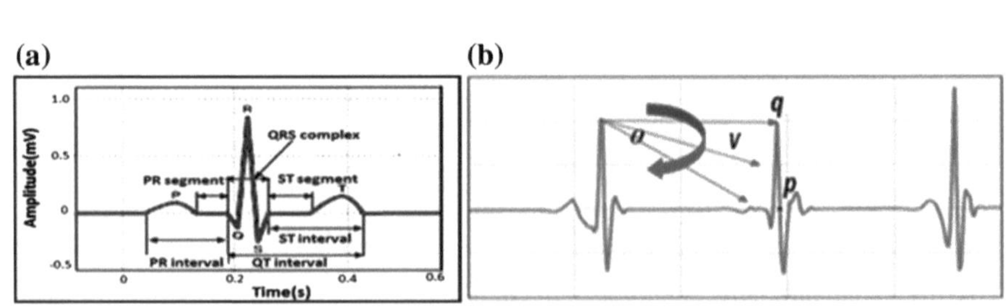

Fig. 3 **a** A typical ECG waveform. **b** Motion of radial vector across P-P complex

$$r' = \frac{(2x(t)x(t) - 2q + 2t - 2p)}{r}$$

$$r[n+1] - r[n] = \frac{2x[n][x[n+1] - x[n]] - 2q}{r(n)} + \frac{2[n-p]}{r[n]} = diff(r)$$

We see from the above expression that $diff(r)$ is higher in the PQRS complex due to the larger value of this term $[r[n+1] - r[n]]$. Again,

$$\frac{ds}{dt} = \frac{d(r\theta)}{dt} = r \cdot \frac{d\theta}{dt} + \theta \cdot \frac{dr}{dt}$$

In discrete time domain, applying the forward difference formula in $\frac{ds}{dt}$,

$$diff(s) = s[n+1] - s[n] = r[n] \cdot [\theta[n+1] - \theta[n]] + \theta[n] \cdot [r[n+1] - r[n]]$$

Thus $diff(s) > diff(\theta)$, since $diff(r)$ is high. In linear traversal, the distance between two consecutive samples is greater than the corresponding distance due to angular rotation of V. Hence, it is faster to traverse by rotation of V across the ECG samples. GBM aims to attain the initial sparsity level in time domain before wavelet decomposition.

Encoding algorithm 1: Geometry-Based Method (GBM)
Input: Filtered ECG column matrix $x[n]$
Initialize:

$$v(counter) = 0,\ \emptyset = 0.5,$$
$$\sigma = \beta, \theta = angular\ displacement\ of\ V, |\Omega| = threshold\ value$$

Set β to any large value such that it does not cross two consecutive R-R interval. Set Ø to 0.5 as usually only R waveform's peak amplitude value is greater than 0.5. $|\Omega|$ is less than the peak amplitudes of P, Q, R, S, and T waveforms.

```
1. Repeat Until v< length of vector x
2.                If x[v] ≥ Ø: Set σ equal to β
3.                         Repeat Until x[v + σ]<Ø
4.      If x[v + σ]≥Ø: Break;
Else: Increment σ by one
5.                         Repeat Until |θ|falls below σ
6.       If x[v + σ - |θ|]<|Ω| : Set
x[v + σ - |θ|] to 0
7.    Increment |θ| by one
8. Set v equal to v + σ
```

Encoding by wavelet transform: After preprocessing through GBM, WTIT extracts the high-frequency components of ECG signal, using wavelet transform. WTIT automatically sets high threshold value for lower level coefficients and vice versa. Hence, the higher frequency components remain intact while high sparsity is attained. In WTIT, a window function of fixed length l' slides over each of the wavelet-transformed matrices. The universal threshold of the transformed signal within the window is calculated as per Eq. 4.

Encoding algorithm 2: Wavelet Transform-based iterative thresholding (WTIT)

Input: $\boldsymbol{\alpha}$ is a column vector containing wavelet transform coefficients of ECG signal of a decomposition level, Y calculates universal threshold as per Eq. 4.

Initialize: $\boldsymbol{\omega} = null\,vector$, $m = 0$, $n = 0$, $l' =$ length of window, μ is iterative threshold, I and y are counter variables.

```
Step I: Repeat Until l' < length of vector α
Set m to m + 1, n to l' − (m − 1) ∗ n and I
                      to 0
Step II: Repeat Until I is <n
Set ω [I] to α [I] and Increment I by
                            one
Step III: Set  μ to Y of ω [I] and Set y to (m − 1) ∗ n
Step IV: Repeat Until y≤I + (m − 1) ∗ n
If α[y] is ≤μ: Set  α [y] to 0 and
                            increment y by one
Set  l' to l' + n
```

We consider l' as an even number. When sliding window is centered at a sample numbered n, the matrix to be thresholded is represented as $\boldsymbol{M}_{j,\boldsymbol{n}}$.

$$\boldsymbol{M}_{j,\boldsymbol{n_a}} = \begin{pmatrix} \boldsymbol{D}_j\left[n_a + \frac{l'}{2}\right] \\ \vdots \\ \vdots \\ \boldsymbol{D}_j\left[n_a - \frac{l'}{2}\right] \end{pmatrix} \tag{9}$$

$$\bar{T}_{j,n_a} = \sigma_{M_{j,n_a}} \sqrt{2\log(l')} \tag{10}$$

$\bar{T}_{j,n_a}$ is the universal threshold of $\boldsymbol{M}_{j,\boldsymbol{n_a}}$ whose iterative threshold is $\tilde{\boldsymbol{M}}_{j,\boldsymbol{n_a}}$.

$$\tilde{\boldsymbol{M}}_{j,\boldsymbol{n_a}}(n) = \boldsymbol{M}_{j,\boldsymbol{n_a}}(n) \;\; iff \left|\boldsymbol{M}_{j,\boldsymbol{n_a}}(n)\right| \geq \bar{T}_{j,n_a} \tag{11}$$

$$\tilde{\boldsymbol{M}}_{j,\boldsymbol{n_a}}(n) = 0\, iff \left|\boldsymbol{M}_{j,\boldsymbol{n_a}}(n)\right| < \bar{T}_{j,n_a} \tag{12}$$

Essentially, we obtain a high amount of sparsity of ECG signal without considerable loss in the high-frequency components through our encoding algorithm.

1.3 Conversion of Sparse Matrix into Transmittable Matrix Using CRHFC

After applying the discrete wavelet transform in the WTIT algorithm, we obtain sparse matrices containing detailed and approximate coefficients. These matrices can further be transformed into matrices containing fewer nonzero elements to each of the coefficient row matrices are converted into square matrices of order $(n_1 \times n_1)$. To convert a row matrix into a square matrix, $\left(n_1^2 - n\right)$ zeroes are appended to each of the row matrix, where

Case 1: $n_1 = \sqrt{n}$, when $n \in Z$, where Z is set of positive integers.
Case 2: $n_1 = [\sqrt{n}] + 1$, when $n \in (R - Z)$, where R is set of positive real numbers.

Algorithm: *Compressed Row Huffman Coding (CRHC)*
Input: The row matrix with n_1^2 elements: [**E**]; let $e_{1,j}$ be the jth element. The square matrix [**C**] is formed from [**E**] in the primary step; let $c_{i,j}$ be its element of ith row and jth column.

$$\left[c_{i,j}\right] = \left[e_{1,\,[\{(i-1)\times n_1\}+j]}\right] \tag{13}$$

Step I: CR: The compressed row storage algorithm [5] compresses each of the square matrices [C] and generates three different row matrices: (i) the element matrix [N]; (ii) the column indicating matrix [IN]: (iii) the row indicating matrix [JN]:
Step II: HC: The element row matrix [N], obtained from Step I, is further compressed using Huffman coding. Combination of Step I and Step II gives CRHC.

1.4 Reconstruction

The reconstruction consists of (i) Huffman decoding, (ii) converse of CR algorithm, (iii) inverse wavelet transforms on the coefficient matrices (Table 1).

Table 1 Comparison among ECG compression and algorithms with our algorithm

Algorithm	PRD	CR
Simultaneous OMP [1]	2.57	7.23
BP [3]	1.66	4
BP [2]	<9	3.44
BP [7]	9	2.5
Our encoding algorithm	2.27	9.4

The time complexity of the procedure is $O(n)$

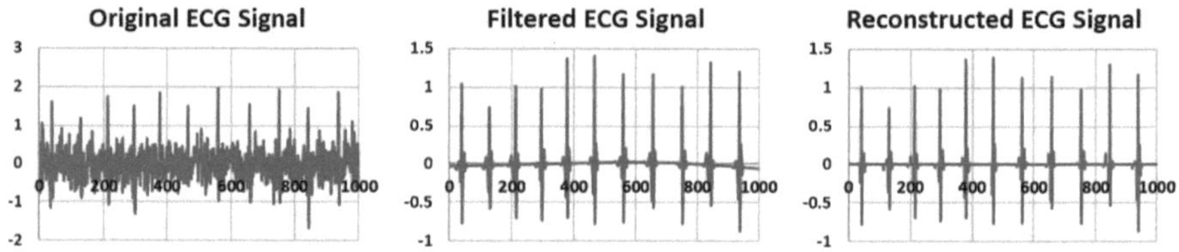

Fig. 4 Performances of our algorithm over an ECG signal producing 91% sparse reconstructed signal

2 Performance Study and Results

To validate the performance of the developed algorithms, we use 15 ECG signal samples from MIT-BIH database [6]. The simulation is performed in MATLAB. The combination of GBM and WTIT achieves a high sparsity level of 87–91%. WTIT and GBM are complementary algorithms. The former does signal processing in time–frequency domain and the later in time domain (Fig. 4).

3 Conclusion

There are several distinct features to our developed scheme namely high sparsity, efficient processing time, ease of data storage, reasonable values of PRD and CR. Our analysis and simulation results suggest that our algorithm is promising for real-time transmission of biomedical signal (in this case ECG) due to its efficient time complexity. Hardware implementation of our algorithm will be explored in our future work.

References

1. Polania L (2011) Compressed sensing based method for ECG compression. In: Proceedings IEEE international conference acoustic, speech signal processes, pp 761–764
2. Mamaghanian H (2011) Compressed sensing for realtime energy-efficient ECG compression on wireless body sensor nodes. IEEE Trans Biomed Eng 58(9):2456–2466
3. Mishra SK, Panda G, Meher S (2009) Chebyshev functional link artificial neural networks for denoising of image corrupted by salt and pepper noise. Int J Recent Trends Eng 1:413–417
4. Hilton ML (1997) Wavelet and wavelet packet compression of electrocardiograms. IEEE Trans Biomed Eng 44:394–402
5. Dongarra J (2000) Sparse matrix storage formats. In: Templates for solution of algebraic eigenvalue problems: a practical guide, SIAM
6. Moody GB, Mark RG MIT-BIH database. http://www.physionet.org/physiobank/database/mitdb
7. Chae DH (2013) Performance study of compressive sampling for ECG signal compression in noisy and varying sparsity acquisition. In: IEEE international conference on acoustic, speech and signal process, pp 1306–1309

Wavelet Based Fractal Analysis of Solar Wind Speed Signal

Tushnik Sarkar, Mofazzal H. Khondekar and Subrata Banerjee

Abstract In the present work, the Solar Wind Speed (SWS) time series has been considered for study. Here, the intention of the authors is to investigate the presence of multifractality within the signal. Continuous wavelet transforms modulus maxima method has been employed in this purpose. The span of the time series of SWS, which is under consideration, is of 2492 days from 01/01/1997 to 28/10/2003. Finally, the Singularity spectrum is drawn to qualitatively evaluate the level of multifractality present within the SWS signal.

Keywords Wavelet transforms · WTMM · Scaling exponent · Hölder exponent
Singularity spectrum

1 Introduction

The stream of ionized plasma and particles that comes out from the outer surface of the sun into the space is known as solar wind. The temperature at the sun's outer layer (corona) is in the range of 2×10^6 °F. The massive temperature increases the kinetic energy (KE) of the particles which reside at the outer surface of the Sun. The large KE makes the particles, around the sun, to overwhelm the gravitational force of the sun and these particles streams out from the sun. But the speed of this solar wind

T. Sarkar (✉)
Department of Electrical Engineering, Dr. B. C. Roy Engineering College,
Durgapur 713206, India
e-mail: tushnik.sarkar@gmail.com

M. H. Khondekar
Department of Applied Electronics and Instrumentation Engineering, Dr. B. C. Roy Engineering College, Durgapur 713206, India
e-mail: hossainkm_1976@yahoo.co.in

S. Banerjee
Department of Electrical Engineering, National Institute of Technology, Durgapur 713209, India
e-mail: bansub2004@yahoo.com

S. Bhattacharyya et al. (eds.), *Recent Trends in Signal and Image Processing*,
Advances in Intelligent Systems and Computing 727,
https://doi.org/10.1007/978-981-10-8863-6_5

is fluctuating in nature. The researchers across the world are yet grasp the absolute understanding of the dynamics of the SWS. Presently it is assumed that the nonlinear processes of the solar corona like MHD turbulence, magnetic reconnection, Alfven wave dissipation etc. are the primary source of the solar wind (particularly slow speed 400 km/s). The findings as obtained from this study, may add some important inputs about the dynamics of the solar wind. The primary reasons for which the scientists are sparing their efforts on the research of the solar wind are its adverse effects on the earth and hence on human civilisation. Some of the happenings caused by the solar wind like the Aurora Borealis, the Aurora Australis at the North Pole and the South Pole respectively, the glowing tail of comets can be noticed by our bare eyes. Another remarkable effect of the solar wind is the dislodgment of the pole of the earth's magnet from its earlier position (close to Murmansk in Russia) to its current position (in the vicinity of Canada). The solar wind is also responsible for geomagnetic storms which destroyed the Hydro Quebec grid, the microchips in computers at Canada in 1989. Due to these damages in grid and computer systems Canadians had to face long power cut and disruption of the Stock market respectively. In 1998, geomagnetic storm smashed the supporting files that are very useful for satellites and stopped the functioning of 45×10^6 pagers. As the Solar wind consists of magnetically charged particles, it plays major role in the formation of disruptive geomagnetic storm on, our planet, the earth.

The main aim of this work is to investigate about the complex behaviour of the dynamics of the solar wind speed fluctuations. In this context, the oscillation of daily averaged SWS (from 01/01/1997 to 28/10/2003), collected from Space Weather Prediction Centre of National Oceanic and Atmospheric Administration (NOAA) has been considered as time series signal for study. The term fractal was introduced by Mandelbrot [1] to characterize a rough, fragmented, crooked or twisted geometric shape. A fractal system has non-integer dimension and the systems which have more than one such fractional dimensions are called multi-fractal system. Fractal objects are geometrical objects which have fractional dimension. Often these objects have structure within a structure which are self similar to each other at different levels of magnification. This fractal structures within the time series signals are assessed by identifying the presence of multifractality and the singularities of the signal.

A multifractal system has complicated, irregular dynamic behaviour. It is inhomogeneous than a monofractal system. The SWS time series signals are nonstationary. To explore the multifractal property within a non-stationary signals, wavelet transform modulus maxima (WTMM) method is suitable because the tool is immune to nonstationarity.

WTMM [2, 3] method is referred as mathematical microscope for detecting the presence of multifractal structure and singularities (that is discontinuities) within the signal. Within a signal various irregular structures or transient phenomena, called singularities, may embedded. These hidden structures tell more about the characteristics of the signal.

Scaling exponent, multifractal scaling exponent and generalized multifractal dimension are three parameters [4] which are evaluated to assess multifractality within the signal quantitatively. These three parameters are represented as

$\tau(q)$, $h(q)$ and $D(q)$ respectively. The presence of singularities under the signal and the intensity of those singularities are also estimated by Hölder exponent (α) and singularity spectrum $f(\alpha)$.

2 Theory

2.1 Continuous Wavelet Transform (CWT)

Wavelet is a wave that exists for a finite duration of time with zero mean. Different Wavelets have different shape and size. The basic Wavelet functions are recognised as mother Wavelets. One mother wavelet $\psi(t)$ can be stretched or compressed by scaling parameter s for obtaining different resolution and this mother wavelet can be translated over time axis by shifting parameter τ for locating the wavelet at desired position on the time axis. The CWT [5] of a data series is performed by taking the convolution between the data sequences and the scaled and shifted version of the mother wavelet. In the meadow of signal analysis, CWT is a widely used mathematical tool. CWT is mathematically described as:

$$W(s,\tau) = \frac{1}{s}\int_{-\infty}^{\infty}\left\{X(t)\cdot\psi\left(\frac{t-\tau}{s}\right)\right\}dt \qquad (1)$$

where $W(s,\tau)$ is the wavelet coefficients, s: scaling parameter, τ: shifting parameter, $X(t)$: time series signal, t: time, $\psi(s,\tau,t)$: mother wavelet, s and τ are real.

The tool Wavelet tells about the time information of the signal with the help of shift parameter (τ) and it discloses different frequency information within the signal from scale parameter (s). In the particular case of singularity detection and multifractality analysis mother wavelet $\psi(t)$ should pose the vanishing moments property which is mathematically represented as:

$$\int t^n\psi(t)dt = 0 \qquad (2)$$

where n (degree of the polynomial) is less than some positive integer m.

A commonly used class of real-valued wavelet which gratifies the Eq. (2) is the successive derivatives of the Gaussian function

$$\psi^{(n)}(t) = \frac{d^n}{dt^n}e^{-(t^2/2)} \qquad (3)$$

The Wavelet Transform of X(t) with $\psi^n(t)$ can be expressed as:

$$W_n(s,\tau) = \frac{1}{s} \cdot \int_{-\infty}^{\infty} \left\{ X(t)\psi^n \left(\frac{t-\tau}{s} \right) \right\} dt$$
$$= s^n \frac{d^n}{dt^n} W_0 \tag{4}$$

Here:

W_n—Wavelet Coefficient for mother wavelet $\psi^n(t)$ and W_0—Wavelet Coefficient for mother wavelet $\psi(t)$.

From Eq. (4),it is found that higher order derivatives will eliminate higher order trends. In the fluctuation of the daily SWS data series of above mention duration may contain trend up to order 2 as estimated by the authors in their earlier work [6]. So the derivative of a Gaussian of order 3 has been selected to eliminate trends. It will not be sensitive for the trends of degree 4 or higher.

$$\psi^3(t) = \frac{d^3}{dx^3} e^{-(t^2/2)} \tag{5}$$

2.2 Wavelet Transforms Modulus Maxima (WTMM)

The CWT representation of a signal is very redundant. WTMM [7] are sufficient to distinguish the singularity of the signal. The magnitude and location of the WTMM are accounted in this context. The WTMMs are found on the scale-time (s, t) plane at a point (s_0, t_0) where $|W(s_0, t_0)|$ is maximum in the vicinity of t_0.Curves are drawn to connect these maxima in the scale-time (s, t) half-plane. These curves are called maxima lines. These maxima lines have a relation with the singularities within the signal. Each singularity is pointed by these maxima lines. WTMM are computed from the coefficient of Continuous wavelet transforms as:

$$\frac{\partial\, |W(s_0, \tau)|}{\partial \tau} = 0 \tag{6}$$

From the coefficients of WTMM the partition function Z can be obtained. Consider for a fixed scale s, $\{u_n(s)\}$ as the position (time) of all local maxima, here n is an integer. The q's power of all these WTMM are summed up, this summation forms the partition function Z [8, 9]:

$$Z(q, s) = \sum_n |W(s, u_n)|^q \tag{7}$$

where q is real, but $q \neq 0$. Here, q is chosen from −10 to +10 and it is incremented by 0.25. The range of scale s is from 2 to 1024 and it is incremented by 1.

2.3 Singularity Detection

Singularity is essentially discontinuity or rapid changes in a time series signal X(t). Singularities [10] are identified by determining the times at which the modulus maxima of the CWT converge at fine scales. This is the key feature of CWT for singularity detection qualitatively. Further, Holder exponent [11–13] is evaluated to quantify the singularity. Holder exponent is measured from the local approximation of the signal X(t) by a polynomial of certain degree. If X(t) is approximated locally by a polynomial of degree m at point t_0 as:

$$\left|X(t) - p_m(t_0)\right| \leq K\left|t - t_0\right|^{\alpha} \tag{8}$$

Polynomial $p_m(t_0)$ is Taylor series expansion of the signal X(t) at t_0 up to order m. $K\left|t - t_0\right|^{\alpha}$ denotes the error in the approximation of the signal X(t) at t_0 and the α is called Holder exponent, a measure of singularity.

To evaluate the Holder exponent α of the signal X(t) the vanishing moments feature [as described in Eq. (2)] of CWT is employed. If the amount of the vanishing moments of the analysing mother wavelet is at least as high as the degree of m, the wavelet coefficient will capture $K\left|t - t_0\right|^{\alpha}$ i.e. the transient occurred at t_0 beyond the local trend in the vicinity of t_0. The relation between wavelet modulus maxima and the Holder exponent for non-oscillating singularities under the signal X(t) across the scales can be described by power law as:

$$|W(s, \tau_0)| \sim s^{\alpha(\tau_0)} \tag{9}$$

From the log-log plot on both sides of the Eq. (9) Hölder exponent α can be estimated. Depending on the value of $\alpha(\tau_0)$ at every τ_0, we can scan the points of irregularity (opposite of regularity) or singularity. Higher the value of $\alpha(\tau_0)$ lesser the strength of the irregularity or singularity and vice versa.

2.4 Multifractal Analysis

The scaling exponent τ(q) is computed from the partition function Z(q, s) at Eq. (7) as:

$$Z(q, s) \sim s^{\tau(q)} \tag{10}$$

Taking the log of the Eq. (10), τ(q) is being estimated for each value of q. From this τ(q) the singularity spectrum f(α) versus α is developed by Legendre Transform as following equations:

$$\alpha(q) = \frac{d\tau(q)}{dq} \tag{11}$$

$$f(\alpha) = q\alpha - \tau(q) \tag{12}$$

When $X(t)$ is homogeneous monofractal, the $\tau(q)$ varies linearly with q. But for non-homogeneous functions where multifractal properties exist, it exhibits non-linear variation. A single humped shape singularity spectrum $f(\alpha)$ is appeared for a multifractal function. The span of Holder exponent α characterizes intermittent fluctuations of the signal $X(t)$. Here α_{min} corresponds for the strongest singularity and α_{max} for the weakest one. In addition to $\tau(q)$—scaling exponent, $\alpha(q)$—Holder exponent and $f(\alpha)$—singularity spectrum as descried earlier multifractality can be noticed from $h(q)$—generalized hurst exponent and $D(q)$—generalized multifractal dimension. They can be computed from $\tau(q)$ as:

$$h(q) = \frac{1 + \tau(q)}{q}, \quad q \neq 0 \tag{13}$$

and

$$D(q) = \frac{\tau(q)}{q - 1} = \frac{qh(q) - 1}{q - 1}, \quad q \neq 1 \tag{14}$$

If $h(q)$ does not depend on q but $D(q)$ shows q dependence then it means that the time series is monofractal. But for multifractal time series both $h(q)$ and $D(q)$ depends on q. The Hurst exponent H is interrelated to the generalised hurst exponent (also known as multifractal scaling exponent) as below:

$$H = h(q = 2) - 1 \tag{15}$$

3 Results

Figure 1 shows the time series of the daily SWS (from 01/01/1997 to 28/10/2003) after smoothing (exponentially) and [14] denoising.

The absolute values of the CWT coefficients $|W(s, \tau)|$ are shown in Fig. 2. Figure 2 is colour coded with 128 different colours where deep brown means $|W(s, \tau)| = 0$ and white indicates max $|W(s, \tau)|$. Vertical axis of the plot is scale and horizontal axis is time. "Wavelet toolbox" of Matlab software is used for plotting.

Figure 3 shows the WT skeleton which is formed by the maxima lines.

Figure 4 displays the plots of different multifractal parameters, singularity spectrum for the SWS time series Signal.

The value of α_{min}, α_{max}, W and $\alpha(f = f_{max}(\alpha))$ are calculated from Fig. 4 for the SWS signal. These values are shown in Table 1.

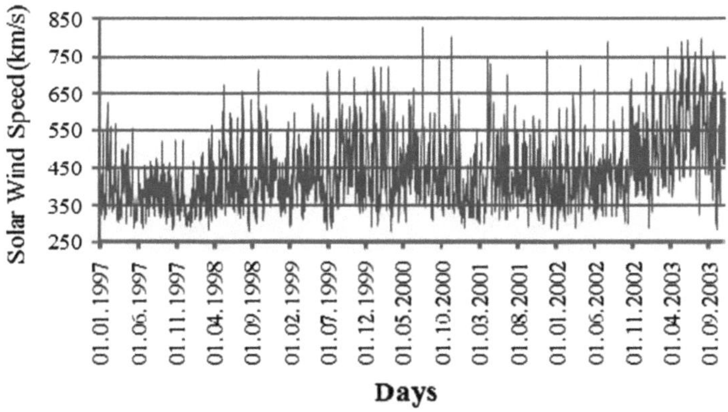

Fig. 1 Solar wind speed time series

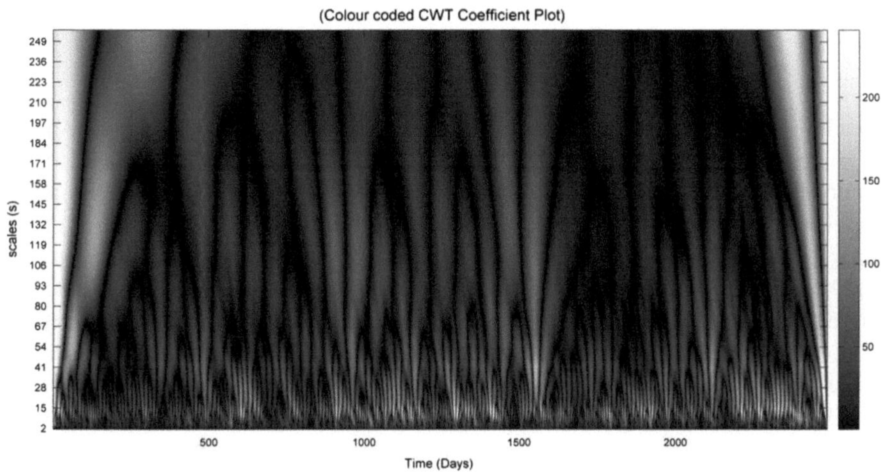

Fig. 2 Diagram (scale vs. time) of colour coded |W(s, τ)| of the SWS signal

4 Discussions

The method WTMM allows us to detect presence of singularity or sharp changes and to characterize multifractality of the fluctuations of the non-stationary daily averaged SWS time series of span 01/01/1997 to 28/10/2003. From Fig. 2 it is observed that the magnitude of CWT of the SWS data series converges from coarser to finer scales at several points. These points of the CWT plot qualitatively indicate those time points where discontinuities have been taken place on time axis. The significant change in slope of the $Z(q)$ with q in Fig. 4a confirms the multifractality of the time series. Figure 4b–e, the nonlinear relations between $h(q)$ versus q, $\tau(q)$ versus q and $\alpha(q)$ versus q respectively, also suggest a multi-fractal nature of the solar wind

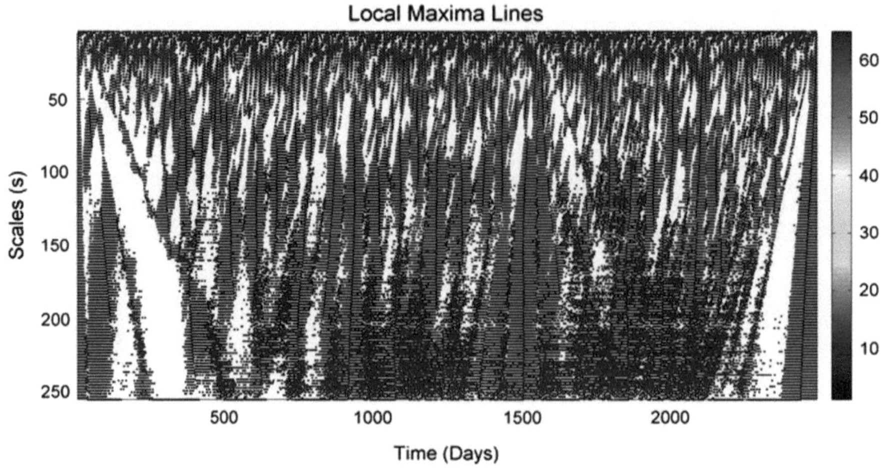

Fig. 3 WTMM skeleton

speed. From Fig. 4b, it can be observed that the value of function $h(q)$ is very much dependent on the order of q. The low value of the $h(q)$ for negative q compared to that for positive values of q in Fig. 4b suggests that the solar wind speed is predominantly populated with large fluctuations. The nature of multifractal characteristics of the time series is more emphasised by portraying the singularity spectrum $f(\alpha)$. In Fig. 4f, singularity spectrum of SWS time series exhibits the shape of a parabolic curve. The important parameters like $\alpha_{\min}$, $\alpha_{\max}$, W & α_0 for SWS data series are computed from Fig. 4f by the curve extrapolation and the results are provided in Table 1. The parameters $\alpha_{\min}$ and $\alpha_{\max}$ are the minimum and maximum value of the Hölder exponent α of the spectrum for which $f(\alpha) = 0$. $f(\alpha)$ determines the amount of occupancy of the signal by the singularities of different strengths. So, $\alpha_{\min}$ and $\alpha_{\max}$ represents the maximum and minimum singularity strengths respectively which are not available in the signal. Thus it can be said that the signal is actually being occupied by the singularities having strengths just above $\alpha_{\min}$ and just below $\alpha_{\max}$. The value of the difference between $\alpha_{\max}$ and $\alpha_{\min}$ i.e. $\mathrm{W} = \alpha_{\max} - \alpha_{\min}$ signifies the width or range of the strength of the singularities (abruptness) available in the signal. The range of fractal exponents in the signal is measured by the spectral width W which indicates the degree of multifractality within the signal. The high value of W in Table 1 indicates that the SWS are rich in fractal structure i.e. it has plentiful of discontinuities or suddenness. This abundant presence of abruptness or irregularity in the SWS signal structure signifies that the information content in the signal is high. The singularities which are most available within the signal are found in the vicinity of α_0 where $f(\alpha)$ is maximum. Since the value of α_0 is high and close to $\alpha_{\max}$, it can be claimed that the mostly available irregularities within the span of W are of low strength.

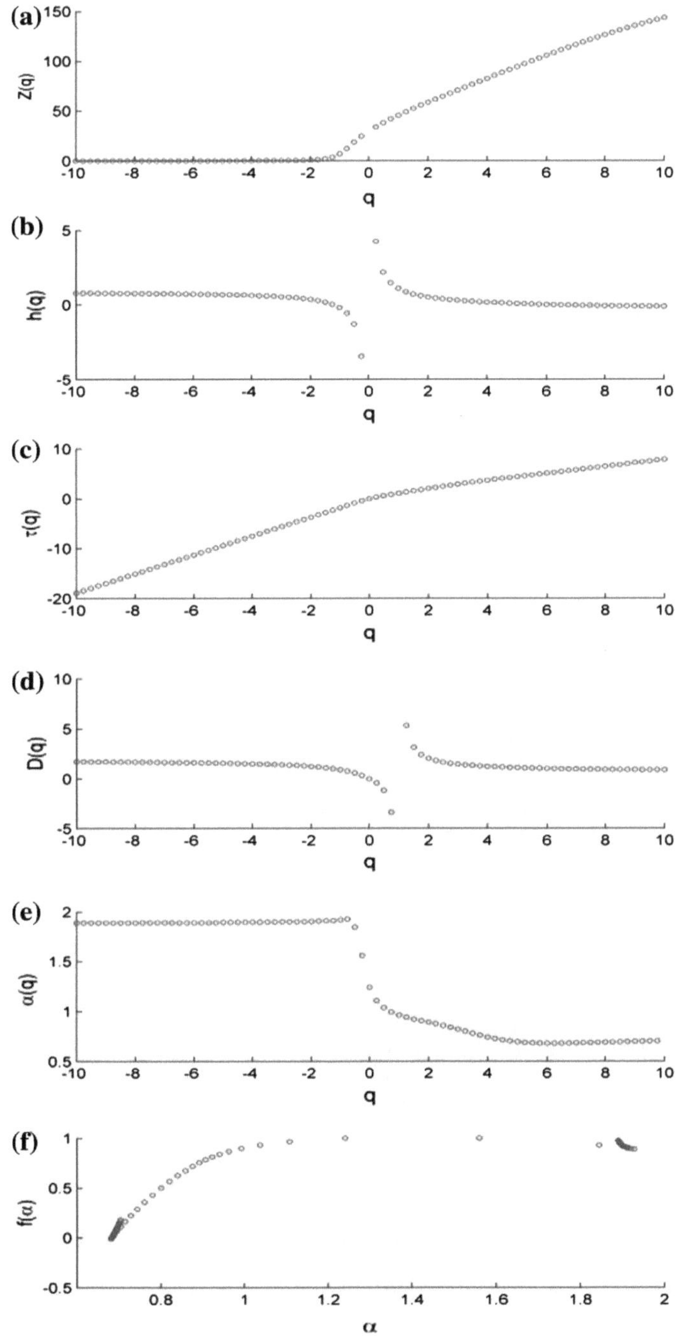

Fig. 4 Plots of different multifractal parameters for the SWS signal: **a** $Z(q)$ against q; **b** $h(q)$ against q; **c** $\tau(q)$ against q; **d** $D(q)$ against q; **e** $\alpha(q)$ against q; **f** $f(\alpha)$ against α

Table 1 Results of α_{min}, α_{max}, W & $\alpha(f = f_{max}(\alpha))$ for the SWS data series

$\alpha_{min}(f = 0)$	$\alpha_{max}(f = 0)$	$W = (\alpha_{max} - \alpha_{min})$	$\alpha(f = f_{max}(\alpha)) = \alpha_0$
0.6811	1.9967	1.3156	1.5605

5 Conclusions

It is realised that whatever the results obtained in the current study on SWS signal by the WTMM tool is in consistence with the earlier work [6] of the authors where MFDFA method was applied. From the above discussions on the obtained results it can be concluded that SWS signal is highly fluctuating, prevailing different fractal structures and singularities of different strengths.

References

1. Mandelbrot BB (1982) The fractal geometry of nature. Freeman, San Francisco
2. Application of the wavelet transform modulus maxima method to T-wave detection in cardiac signals. http://www.personeel.unimaas.nl/westra/PhDMaBaeaching/GraduationStudents/PJouck2004/PJouck2004verslag.pdf
3. Valsamma KM (2013) Multifractal analysis of annual run-off: using WTMM method. Int J Eng Sci Innov Technol (IJESIT) 2(1):139–145
4. Hossain KM, Ghosh DN, Ghosh K (2009) Investigating multifractality of solar irradiance data through wavelet based multifractal spectral analysis. Signal Process Int J (SPIJ) 3(4):84–94
5. Mallat S (1998) A wavelet tour of signal processing. Academic Press, New York
6. Sarkar T, Hossain KM, Banerjee S (2017) Signal processing approach to study multifractality and singularity of solar wind speed time serie. WASET, Int J Comput Electric Autom Control Inf Eng 11(2):97–102
7. Gospodinova E, Gospodinov M (2014) Wavelet-based multifractal analysis of RR time series. Int J Adv Res Comput Sci Softw Eng 4(4):1067–1071
8. Bogdan E, Kiyoshi I, Zbigniew RS (2004) Wavelet-based multifractal analysis of real and simulated time series of earthquakes. Geophys J Int 164(6):63–74
9. Hossain KM, Ghosh DN, Ghosh K, Bhattacharya AK (2011) Multifractality and singularity of 8B solar neutrino flux signals from sudbury neutrino observatory. IET Signal Proc 5(7):690–700
10. Cheng J, Chen D, Landry Jr R, Zhao L, Guan D (2014) Research on wavelet singularity detection based fault-tolerant federated filtering algorithm for INS/GPS/DVL integrated navigation system. J Appl Math 2014:1–8
11. Mallat S, Hwang W (1992) Singularity detection and processing with wavelets. IEEE Trans Info Theory 38:617–643
12. Mallat S, Zhong S (1992) Characterization of signals from multiscale edges. IEEE Trans Patt Recog Mach Intell 14:710–732
13. Mathuvanesan C, Jayasankar T (2013) Performance analysis of singularity and irregular detection in human health monitoring using Lipschitz exponent function. Int J Eng Res Technol 2(6):414–418
14. Sarkar T, Ray R, Khondekar MH, Ghosh K, Banerjee S (2015) Chaos and periodicity in solar wind speed: cycle 23. Astrophys Space Sci 357(2):1–10

Class Discriminator-Based EMG Classification Approach for Detection of Neuromuscular Diseases Using Discriminator-Dependent Decision Rule (D3R) Approach

Avik Bhattacharya, Purbanka Pahari, Piyali Basak and Anasua Sarkar

Abstract Classification of EMG signals is essential for diagnosis of motor neuron diseases like neuropathy and myopathy. Although a number of strategies have been implemented for classification, none of them are efficient enough to be implemented in clinical environment. In the present study, we use ensemble approach of support vector machines for classification of three classes (normal, myopathic and neuropathic) of clinical electromyogram (EMG). Our proposed approach uses time and time–frequency features extracted from EMG signals. By employing two types of feature set for same class discriminators, we are able to select the best feature set-discriminator pairs. The decision made by each selected classifier is used to generate the final class for an input EMG signal through majority voting. Our proposed method yields higher accuracy of 94.67% over 89.67% for multiclass SVM classifier.

Keywords Ensemble framework · Discriminator-dependent decision rule Electromyogram · MUAP classification

1 Introduction

During muscular activity, muscle fibres produce train of electrical discharges which can be identified as discrete waveforms known as motor unit action potential (MUAP). During increased muscular activities, it becomes practically impossible to

A. Bhattacharya · P. Pahari · P. Basak
School of Bioscience and Engineering, Jadavpur University, Kolkata, India
e-mail: avik94ab@gmail.com

P. Pahari
e-mail: purbankapahari36@gmail.com

P. Basak
e-mail: dir_bioengg@school.jdvu.ac.in

A. Sarkar (✉)
Computer Science and Engineering Department, Jadavpur University, Kolkata, India
e-mail: anasua.sarkar@cse.jdvu.ac.in; ashru2006@hotmail.com

S. Bhattacharyya et al. (eds.), *Recent Trends in Signal and Image Processing*,
Advances in Intelligent Systems and Computing 727,
https://doi.org/10.1007/978-981-10-8863-6_6

identify individual MUAPs separately. As a result, it is not feasible to extract enough quantitative information from the complex MUAP firing patterns for diagnosis of neuromuscular disease. Hence, there is a need to use soft computing techniques to decompose the complex interference patterns of MUAPs into identifiable waveforms for further quantitative analysis [1].

Subasi [2] proposes a method to classify clinical EMG signals using time–frequency features as input to ANN, ANFIS and DFNN classifiers. Subasi [3] uses a combination of particle swarm optimisation (PSO)-based SVM to classify myoelectric signals on the basis of neuromuscular diseases. Kamali et al. [4] explore nine morphological time and time–frequency-domain features (DWT) of clinical myopulses and then perform the final classification by using ensemble framework. However, the study uses only the best classifier from the one-versus-all classifier pool instead of selecting all the three best classifiers. Consequently, the approach is prone to single error.

In our ensemble framework, discriminator-dependent decision rule (D3R) generates final class labels for MUAPs through a classification process that involves multiple feature sets (time and time–frequency) and class discriminators. The approach is primarily based on majority voting among multiple class discriminators. This classification scheme is not single error-prone and yields higher accuracies compared to other multiclass SVM classifiers using the same feature sets (time and time–frequency) and combination of both.

2 Methodology

In this section, we discuss the dataset and the feature sets used for our present work to quantitatively analyse MUAPs.

2.1 Dataset Description

We use 100 clinical EMG sequences corresponding to each class (normal, neuropathic and myopathic), with total of 300 EMG signals. The signals are digitized at 23437.5 Hz and contain 4096 EMG samples. The signals are collected from EMGLAB [5].

2.2 Feature Extraction and Reduction

2.2.1 Time-Domain Features (T)

Time (T) features are evaluated considering the EMG signal to be stationary in nature. The following time (T)-domain features are extracted in our work [6, 7].

1. Mean Absolute Value (MAV):MAV is defined as the average of absolute values of all EMG samples in a sequence. Mathematically,

$$MAV = \frac{1}{n}\sum_{i=1}^{n} | x_i |$$

   where x_i is the amplitude of ith EMG sample and n is the sequence length.
2. Root Mean Square (RMS): RMS value provides an indication of the energy content in a given EMG sequence. Lower energy levels is a characteristic feature of diseased EMG signals. It is defined as,

$$RMS = \sqrt{\frac{1}{n}\sum_{i=0}^{n} x_i^2}$$

3. Zero Crossing (ZC): ZC provides information related to frequency but is defined in time domain. It is a counter indicating the number of times the EMG signal crosses the zero-amplitude axis. Mathematically,

$$ZC = \sum_{i=1}^{n-1}[sgn(x_i \times x_{i+1}) \cap | x_i - x_{i+1} | \geq threshold]$$

   where $sgn(x) = 1 \; if \; x \geq threshold$ and $= 0$ otherwise.
4. Waveform Length (WL): The cumulative length of an EMG waveform is defined as wavelength (WAVE/WL), which indicates the complexity of EMG signals. WL is defined as,

$$WL = \sum_{i=1}^{n-1} | x_{i+1} - x_i | .$$

2.2.2 Time–Frequency-Domain Features (TF)

DWT analysis occurs by decomposing the signal at varying frequency bands and resolutions [8]. The process of decomposition occurs by feeding a discrete-time signal $x[n]$ through half-band low-pass filters and high-pass filters defined by transfer functions $g[n]$ and $h[n]$, respectively. We express the decomposition as:

$$y_{high} = \sum^{n} x[n].g[2k - n]$$

and

$$y_{low} = \sum^{n} x[n].h[2k - n]$$

where y_{high} and y_{low} are the outputs of low- and high-pass filters at a particular level of decomposition, respectively. In this study, we use Symlet4 mother wavelet at the fifth level of decomposition as this level provides the best possible discrimination among clinical classes of EMG [4]. We reduce the dimensionality of the initial feature vector by evaluating three statistical features, namely average of absolute values, standard deviations and average powers of the DWT coefficients of each of the sub-bands and the ratio of means of DWT coefficients of adjacent sub-bands [3].

1. Average of absolute values of DWT coefficients of each of the sub-bands
2. Standard deviation of the DWT coefficients of each of the sub-band
3. Average power of the DWT coefficients of each of the sub-bands
4. Ratio of mean of the DWT coefficients in the adjacent sub-bands

3 Framework for Classification

A total of twelve classifiers are generated using six class discriminators. The classifiers are formed by using either time or time–frequency features as shown in Table 1. Class discriminators D_1 to D_6 are designated as one against rest (OAR), while D_7 to D_{12} are designated as one against another (OAA). Linear support vector machine is used as the preliminary model to train and tenfold cross-validate the classification models [9].

Once the validation accuracies of all the single classifiers are available, we select the most accurate classifier out of the pool of twelve classifiers. Once a classifier

Table 1 List of classifiers

Discriminator	Class discrimination	Feature set	Discriminator	Class discrimination	Feature set
D_1	Normal versus others	T	D_2	Normal versus others	TF
D_3	Myopathy versus others	T	D_4	Myopathy versus others	TF
D_5	Neuropathy versus others	T	D_6	Neuropathy versus others	TF
D_7	Normal versus myopathy	T	D_8	Normal versus myopathy	TF
D_9	Myopathy versus neuropathy	T	D_{10}	Myopathy versus neuropathy	TF
D_{11}	Normal versus neuropathy	T	D_{12}	Normal versus neuropathy	TF

Table 2 Permissible classifiers selection for decision rule generation

MUAP type	D_1	D_2	D_3	D_4	D_5	D_6	D_7	D_8	D_9	D_{10}	D_{11}	D_{12}
Normal	1	1	1	1	1	1	1	1	0	0	1	1
Myopathic	1	1	1	1	1	1	1	1	1	1	0	0
Neuropathic	1	1	1	1	1	1	0	0	1	1	1	1

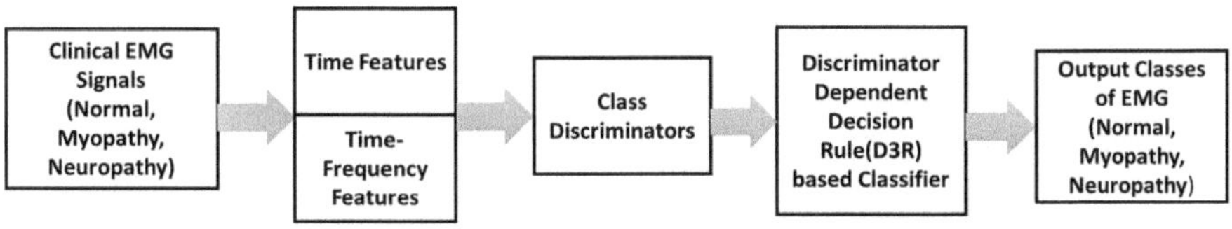

Fig. 1 Flowchart of the proposed ensemble framework to diagnose diseased MUAPs

is selected, the classifier performing similar class discrimination is removed from the pool. For example, if D_8 is selected, then D_7 is automatically removed from the pool as both of them are responsible for separating normal class from myopathic class. This process is continued until we end up with six classifiers (one each for every class discriminator). This ensures that we have the best discriminator-feature combinations in our final pool of six classifiers.

A *permissible classifier* is capable of contributing successfully to the recognition of a particular class of clinical MUAP. For example, D_{12} is a permissible classifier for classifying healthy and neuropathic EMG signals and not myopathic ones. So, for each input class, there exist five permissible classifiers. The classifiers that can recognise a given class are assigned '1', while '0' is assigned if class cannot be recognised. The list of permissible classifiers for each MUAP class is shown in Table 2. Depending upon the input class label of MUAP, five permissible classifiers are selected out of the six best classifiers elected. The 'other' class discriminator is treated by voting both other classes simultaneously. The final output class for each EMG sequence is assessed by majority voting. The class with most votes is selected as the final class. Cases where two or more classes have the same votes is addressed by counting the number of votes by the OAR classifiers against each class and selecting the class with maximum votes as the output class for a given EMG. The model is graphically shown in Fig. 1.

4 Results

For analysing the accuracy of different classifiers and discriminators used in the work, we compare them with gold standard class labels assigned to the EMG signals by clinicians.

In case of single classifiers, we observe that time features yield higher accuracies when discriminating myopathy from others and myopathy from neuropathy. For all other class discriminators, the classifiers with time–frequency-domain features provide higher accuracies in comparison with the classifiers with time-domain features. As shown in Table 3, it is obvious that we select D_2, D_4, D_5, D_8, D_9 and D_{12} classifiers for decision rule generation in our proposed model. Using the final ensemble framework of our new approach, we obtain results that are highly accurate and perfect for implementation in clinical settings. The high sensitivity and specificity values as obtained by all the classes in D3R approach signify the utility of our classification scheme proposed classification to efficiently detect diseased myopathic and neuropathic MUAPs in EMG signals.

We compare the results of our ensemble framework with conventional multiclass SVM approach that employs time features, time–frequency features and a combination of both (SVM_T, SVM_{TF}, SVM_{T-TF}) as shown in Table 4. The comparative analysis is based upon these indicators: the overall accuracy (Acc) and sensitivity (Sen_{Myo}, Sen_{Neuro}, Sen_{Normal}) and specificity (Spc_{Myo}, Spc_{Neuro}, Spc_{Normal}) value for the three classes of clinical EMG. Figure 2 shows the comparison graphically. Our proposed D3R approach of classification yields the highest accuracy of classifier compared to the other three methods analysed in our study as shown in Fig. 2. The other methods are developed using a multiclass SVM with different feature sets (T,TF and T-TF). The sensitivity and specificity values for all clinical conditions as obtained by our approach are higher than the other approaches. The other methods have lower values for sensitivity and specificity for at least some of the clinical

Table 3 Classification accuracy of the class discriminators

Discriminator	Accuracy (%)	Discriminator	Accuracy (%)
D_1	77.33	D_2	88.67
D_3	83.33	D_4	86.67
D_5	97.33	D_6	96.67
D_7	96.50	D_8	99.50
D_9	98.50	D_{10}	97.50
D_{11}	75.50	D_{12}	87.00

Table 4 Performance Comparison of D3R with other approaches

Classifier	Acc. (%)	Spc_{Myo} (%)	Spc_{Neuro} (%)	Spc_{Normal} (%)	Sen_{Myo} (%)	Sen_{Neuro} (%)	Sen_{Normal} (%)
SVM_T	81	96	87	89	98	78	68
SVM_{TF}	89.67	97.5	92.5	94.5	98	90	81
SVM_{T-TF}	89.67	98.5	91.5	94.5	99	81	81
$D3R$	94.67	99	96	97	99	94	90

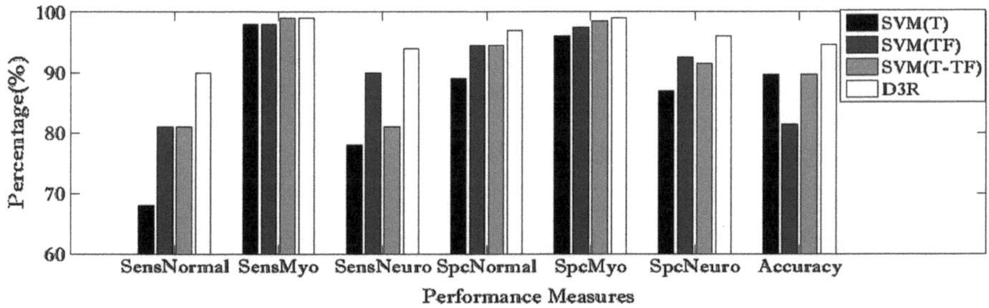

Fig. 2 Comparison of performance measures of classification approaches

conditions. D3R approach provides an accuracy of 94.67% compared to the other methods which yield below 90% accuracy values on our chosen dataset.

5 Conclusion and Future Scopes

In this work, we propose the discriminator-dependent decision rule (D3R)-based new classifier using ensemble framework for diseased MUAP detection, with 94.67% classification accuracy. On the other hand, the multiclass SVM-based classifiers provide accuracies of 81, 89.67 and 89.67% when time, time–frequency and ensemble of both features are employed, respectively. It also becomes clear from the framework of our study that we select the best feature-discriminator combination. This feature optimisation technique through D3R approach enhances the accuracy by a large margin.

Scopes for future development of the proposed model can be through implementation of advanced classifiers instead of linear SVM. In order to remove the effect of noise on the classification performance, frequency features can be included as part of the classification scheme.

References

1. Merletti R, Parker P (2004) Electromyography: physiology, engineering, and non-invasive applications. Wiley, Hobokken, NJ
2. Subasi A (2012) Classification of EMG signals using combined features and soft computing techniques. Appl Soft Comput 12:2188–2198
3. Subasi A (2013) Classification of EMG signals using PSO optimized SVM for diagnosis of neuromuscular disorders. Comput Biol Med 43:576–586
4. Kamali T, Boostani R, Parsaei H (2014) a multi-classifier approach to MUAP classification for diagnosis of neuromuscular disorders. IEEE Trans Neural Syst Rehab Eng 22(1), 191–200
5. Hamilton WA, Stashuk DW (2005) Physiologically based simulation of clinical EMG signals [dataset S003]. IEEE Trans Biomed Eng 52, 171–183. www.emglab.net

6. Elamvazuthi I, Duy HXN, Ali Z, Su WS, Khan A, Parasuraman S (2015) Electromyography (EMG) based classification of neuromuscular disorders using multi-layer perceptron. Procedia Comput Sci 76:223–228
7. Phinyomark A, Phukpattaranont P, Limsakul C (2012) Feature reduction and selection for EMG signal classification. Expert Syst Appl 39:7420–7431
8. Akansu Ali N, Haddad Richard A (1992) Multiresolution signal decomposition: transforms, subbands, and wavelets. Academic Press, Boston, MA
9. Cortes C, Vapnik V (1995) Support-vector networks. Mach Learn 20(3):273–297

A Cooperative Co-evolutionary Approach for Multi-objective Optimization

Sharbari Basu, Ankur Mondal and Abhishek Basu

Abstract In today's world, the interest on multi-objective optimization through evolutionary algorithms (EAs) is growing day by day. However, most of the relative researches are confined within small scale with relatively fewer number of decision variables limited within 30, though real-world multi-objective optimization problems deal with, most of the times, with more than hundred decision variables. Also, optimization with fully separable decision variables along with non-separable decision variables leads to more optimal solutions than dealing with only any one of the both. In this paper, we have proposed an algorithm, which deals with medium- to large-scale multi-objective decision variables, compares the optimal solutions of separable and non-separable decision variables, and accepts the one having most optimized decision. Here we have adopted the test functions (large-scale multiobjective and many-objective optimization test problems for separable decision variables and Zitzler–Deb–Thiele test suit for non-separable decision variables) that are scalable with more than 100 decision variables and can range the results of both separable and non-separable decision variables.

Keywords Inverted generational distance (IGD) · Large-scale multi-objective and many-objective optimization test problems (LSMOPs) · Multi-objective optimization differential evolution (MODE) · Multi-objective optimization problems (MOPs) · Multi-objective optimization evolutionary algorithms (MOEAs) · Non-separability · Pareto optimal set · Separability Zitzler–Deb–Thiele test suit (ZDT)

S. Basu (✉) · A. Mondal
Department of Computer Science and Engineering, GNIT, Sodepur, Kolkata 700 114, India
e-mail: basu.sharbari@gmail.com

A. Mondal
e-mail: mondal.ankur@gmail.com

A. Basu
Electronics and Communication Engineering Department, RCCIIT, Canal South Road, Beliaghata, Kolkata 700015, India
e-mail: idabhishek23@yahoo.com

S. Bhattacharyya et al. (eds.), *Recent Trends in Signal and Image Processing*,
Advances in Intelligent Systems and Computing 727,
https://doi.org/10.1007/978-981-10-8863-6_7

1 Introduction

The real world demands optimization of more than a single objective function simultaneously all at once. Those problems are named as multi-objective optimization problems (MOPs). Their solutions involve best possible balancing between the two opposing functions, i.e., those are conflict with each other. Thus, it involves a solution set termed *Pareto optimal set* [1], and value of the corresponding objective function values form the *Pareto front.*

Since many years, MOPs are being tried to solve through mathematical techniques [2], but as in fact, the real-world applications of MOPs are mostly nonlinear, even sometime non-differentiable. This has increased the importance of metaheuristic approaches, and among these approaches, evolutionary algorithms (EAs) have gained its own importance. Multi-objective evolutionary algorithms (MOEAs) have a big advantage; that is, it is population based, resulting generation of several Pareto optimal set [1] elements in a singular run, while the mathematical models produce single element per run. As per this very paper is concerned, we have used multi-objective differential evolution (MODE) to solve the MOPs.

In this projected approach, we have dealt with more than 100 decision variables as in the real world, though most of the metaheuristic approaches ponder around less number of decision variables (in most cases, not more than 30) and there is empirical evidence that most of these presently available approaches have significantly decreased the efficacy, on increasing the decision variables number in MOPs [3, 4]. In this work, we have proposed a cooperative co-evolutionary framework, that allows MODE to handle extensive number of decision variables, which splits large populations, based on separability and non-separability of decision variables, and then accordingly finds the Pareto optimal set [1]. One of the chief reasons behind using a cooperative co-evolutionary framework, as it is evident that such framework is very much effective in handling large-scale global optimization problems [1].

2 Basic Concepts

During multi-objective optimization problems, we failed to get a definite solution as they involve objectives which are in conflict with each other. So here we get Pareto optimal set. MODE utilizes Pareto-based assignment with crowding distance metric.

However, we focus on solving problems of type:
minimize

$$\vec{f}(\vec{x}) = [f_1(\vec{x}),\ f_2(\vec{x}),\ \ldots,\ f_k(\vec{x})]^T \tag{1}$$

subject to:

$$g_i(\vec{x}) \leq 0 \quad i = 1, 2 \ldots m \tag{2}$$

$$h_i(\vec{x}) = 0 \quad i = 1, 2 \ldots p \tag{3}$$

k symbolizes the number of objective functions f_i: $\mathbb{R}^n \cdot \mathbb{R}$ [5],
g_i, h_j: $\mathbb{R}^n \cdot \mathbb{R}$, $i = 1, \ldots, m$, $j = 1, \ldots, p$ symbolizes constraint functions of the given problem; moreover, $\vec{x} = [x_1, x_2, \ldots, x_n]^T$ defines the vector of decision variables [1]. So, our objective is to determine from set Ω (where Ω is the feasible region) of all the vectors which satisfies both Eqs. (2) and (3) to the vector $\vec{x} = [x_1^*, x_2^*, \ldots, x_n^*,]^T$ which are *Pareto optimal* [1].

To define optimality theory in which we are interested, here few definitions are being quoted [6, 7]:

Definition 1 It is said that **x dominates y (notation** x $\prec$ y) if x $\le$ y and x $\ne$ y when, two given vectors x, y $\in \mathbb{R}^k$, we state that

$$\mathrm{x} \le \mathrm{y} \quad \text{if}\, x_i \le y_i \;\text{for}\, i = 1, \ldots k \tag{4}$$

Definition 2 It is said that a vector of decision variables x $\in \chi \subset \mathbb{R}^n$ is **non-dominated** respecting χ, if there does not exist another x'$\in \chi$ such that *f(x′) $\prec$ f(x).*

Definition 3 It is said that a vector of decision variables

$$x* \in F \subset \mathbb{R}^n \tag{5}$$

(where *F* is feasible region) is **Pareto optimal** if it is non-dominated w.r.t *F.*

Definition 4 The **Pareto optimal set** P^* [1] can be stated as

$$P^* = \{x \in F \,|\, x \text{ is Pareto Optimal}\} \tag{6}$$

Definition 5 The **Pareto front** can be predicted as

$$PF^* = \{f(x) \in \mathbb{R}^k \,|\, x \in P^*\} \tag{7}$$

3 Multi-objective Differential Evolution

On extending differential evolution algorithm on multi-objective optimization problems, two key aspects are being considered:

(i) the way to apply diversity into the population.
(ii) the way to execute *elitism*.

Here, below, authors briefly define the above-mentioned design aspects.

(a) **Applying diversity**: This is done through **crowding distance** factor and through **fitness sharing**.

Crowding distance [8]—This factor gives idea that how a given individual is being crowded by its closest neighbors in an objective function space. Individual with larger value of this factor is being preferred.

Fitness sharing [9, 10]—An individual's fitness is degraded when it shares its resources with others, in percentage to the number and closeness to it which surrounds it within a definite boundary. The locality of the individual is defined by σ_{share} indicating neighborhood radius. These neighborhoods are termed *niches.* Preference given to an individual whose *niches* is less crowded.

(b) **Performing Elitism**: Elitism in evolutionary multi-objective optimization is realized through secondary population that stores non-dominated individuals found along the search. The most popular method is *non-dominated sorting approach* [11].

4 Interacting and Non-interacting Genes

There lies a high probability of falling two interacting variables or non-separable genes in the same subcomponent, after being grouped randomly at commencement of every cycle.Below equation calculates the probability of grouping v interrelating variables within the very subcomponent v interacting variables in the same subcomponent for no less than k cycles [12].

$$\Pr(X \geq k) = \sum_{r=k}^{N} \binom{\mathrm{N}}{\mathrm{r}} \left(\frac{1}{\mathrm{m}\mathrm{v}-1}\right)^{r} \left(1 - \frac{1}{\mathrm{m}^{\mathrm{v}-1}}\right)^{N-r} \tag{8}$$

where Pr = probability of assigning variables $x_1, x_2, \ldots, x_v$ in one subcomponent for at least k cycles; N = number of cycles; v = number of variables; m = number of subcomponents; X = number of times, while v interacting variables are clustered into single subcomponent, where $X \geq k, k \leq N$.

On the basis of the above equation, we can calculate the probability of interacting genes. So non-interacting or separable genes are

$$Pr' = 1 - Pr \tag{9}$$

5 Proposed Approach: Grouping Framework

In this script, authors have used cooperative coevolution (CC) to split the population into subpopulations. CC utilizes a divide-and-conquer practice which splits the decision variables into trivial size subpopulations and those subgroups are being optimized by using EAs. To get the desired result, we have used *MOEA/D-DE* [13].

Table 1 Comparison of non-dominated solutions of MOEA/D (s&ns) with NSGA-II on magnitude 2

Problem	Values	MOEA/D (s&ns)	NSGA-II
LSMOP1	Best	**0.0189**	0.305
	Median	**0.0267**	0.313
	Worst	**0.0686**	0.323
LSMOP5	Best	0.0135	0.341
	Median	0.0154	0.342
	Worst	0.0181	0.344
LSMOP9	Best	0.316	0.811
	Median	**0.336**	0.811
	Worst	**0.350**	0.811

As per our proposed work, at the very beginning, the main population *NP* having decision variables $\vec{x}$ of dimension $D \in$ N is being splitted into two basic groups G_i and G_j based on the separability criterion, where G_i holds 'separable genes' and G_j holds "non-separable genes." If number of decision variables $\vec{x}_i = \{\vec{x}_{i1} \ldots \vec{x}_{in}\}$ of G_i or $\vec{x}_j = \{\vec{x}_{j1} \ldots \vec{x}_{jn}\}$ of G_j is less than or equal to half of *NP*, then multi-objective differential evolution (*MOEA/D-DE*) is being applied on each group. But, if $\vec{x}_i$ and $\vec{x}_j$ are greater than half of *NP*, then two different methods are being applied. In case of $G_{i,}$ the group is further divided into subgroups and again their sizes are compared with the half of size of *NP*. Then, *MOEA/D-DE* is applied on each subgroup. The most non-dominating decision variables are being selected. On the other hand, in case of G_j, the crowding distance and the dominance factor of each gene are being evaluated. Accordingly, groups are formed whose sizes must be less than or equal to half of *NP*. On each subgroup, *MOEA/D-DE* is being applied and then the most non-dominating decision variables are being selected. Finally, comparison between the best non-dominating variables from G_i and G_j is being selected resulting the *Pareto optimal set* [1]. The proposed work is being named as MOEA/D (s&ns), i.e., Multiobjective Evolutionary Algorithm based on Differential Evolution (with separable and non-separable decision variables) (Table 1).

The authors are about to apply it on multi-objective environment of **Digital Watermarking**.

Here we have adopted the test functions large-scale multi-objective and many-objective optimization test problems (LSMOPs) [14] for separable decision variables and Zitzler–Deb–Thiele test suit (ZDT) [15] for non-separable decision variables. To measure the result set quality by means of MOEA, the inverted generational distance (IGD) [16–18] be employed as a performance indicator.

$$IGD\left(P^{*},\ P\right) = \frac{\sum\limits_{v \in P*} d(v,\ P)}{|P*|} \tag{10}$$

where

P^* is set of homogeneously distributed solutions on the proper Pareto font,
P is set of approximate solutions acquired by an MOEA
$d(v, P)$ is minimum Euclidian distance among a point v from P^* and every point in P

6 Algorithm

```
Require: NP, Cycles, D
          Ensure:  Set of Solutions
           Check ←Separable genes(NP)
             Gi← Check
            Gj← NP-Check
            if (Gi≤NP/2) || (Gj≤ NP/2)
            {
                MOEA/D-DE(Gi )
               MOEA/D-DE (Gj)
              }
              else if  (Gi>NP/2) || (Gj>NP/2)
                {
          crd ← crowding distance(Gj)
        dom←dominance (Gj)
                for (j=1;j≤NP;j++)
{
                 MOEA/D-DE(Gj)
        y← non-dominating gene(Gj)
}
                 p ← NP/2
                 for (Gi≤p; Gi=0; Gi--)
               {
                   MOEA/D-DE(Gi)
                 z←non-dominating gene(Gi)
                }
               p=p+NP/2
                 }
                 Set of Solutions ←best(compare(y,z))
                     return  Set of Solutions
```

7 Analysis of Results

Tables 2 and 3 hold the data of both non-separable and separable non-dominated solutions, respectively. The experiment is being done on 200 and 300 decision variables with around 2000 and 3000 generations. So the result tends to be more accurate. The best, median, and worst IGD values in both cases are being considered, and finally

Table 2 Evaluations of non-separable genes (G_j) on ZDT1, ZDT2, ZDT3. Best, median, and worst IGD values obtained on bi-objective and three objective instances

No. of decision variables	Max generation used	Magnitudes	Best	Median	Worst
ZDT1					
200	2000	2	0.003898	0.004052	0.00420
300	3000	3	0.003786	0.004012	0.004100
ZDT2					
200	2000	2	0.003694	0.003715	0.003813
300	3000	3	0.003518	0.003681	0.003742
ZDT3					
200	2000	2	0.007048	0.007079	0.007184
300	3000	3	0.007046	0.007072	0.007176

Table 3 Evaluations of separable genes (G_i) on LSMOP1, LSMOP5, LSMOP9 [14]. Best, median, and worst IGD values obtained on bi-objective and three objective instances

No of decision variables	Max generation used	Magnitudes	Best	Median	Worst
LSMOP1					
200	2000	2	0.0189	0.0267	0.0686
300	3000	3	1.01	1.1	1.16
LSMOP5					
200	2000	2	0.0135	0.0154	0.0181
300	3000	3	0.601	0.7	1.208
LSMOP9					
200	2000	2	0.316	0.336	0.35
300	3000	3	0.413	0.466	0.484

they are compared. The comparison between two tables is being carried firstly on best values, then on median values, and finally on worst values. As we can see from the above tables, the non-separable genes perform better than separable genes on both magnitudes 2 and 3. So overall, we can conclude, that the IGD values are more perfect in case of separable genes (Fig. 1).

8 Comparison

In this paper, we have compared our non-dominated solutions (separable genes) with NSGA-II [19] on magnitude 2. We can clearly conclude, our proposed algorithm performs far superior than that of NSGA-II. NSGA-II is a multi-objective GA. In

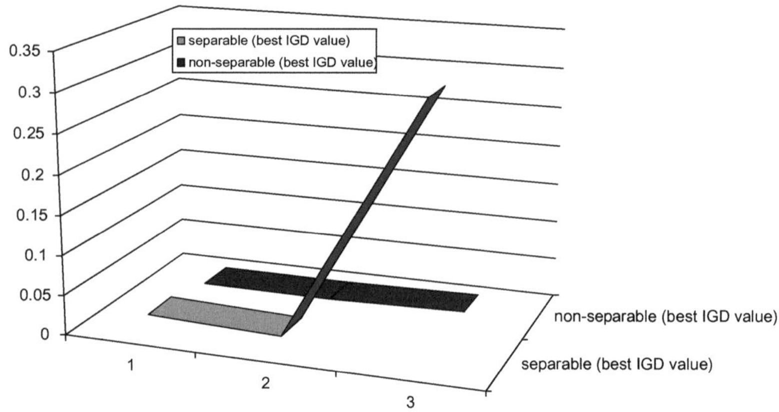

Fig. 1 Performance comparison between best IGD values of separable and non-separable genes on magnitude 2

future work, the authors are going to compare with the many-objective evolutionary algorithm as proposed by Prof. Ishibuchi.

9 Future Scope and Conclusion

In this paper, we have proposed MOEA/D (s&ns) which can simultaneously give non-dominated solutions of both separable and non-separable decision variables and give the finest non-dominated solutions.

The authors of this paper are about to proceed on the application of MOEA/D (s&ns) in "**Digital Watermarking.**" The three criteria of "**Digital Watermarking,**" namely *robustness, hiding capacity, and imperceptibility* are inversely related to each other, i.e., if one gives optimized result, the results of the other two degrades. MOEA/D (s&ns) can be applied on this multi-objective environment so as to get the desired result.

This paper can also be utilized for image processing, digitization of image, and also on network planning.

References

1. Luis Miguel A, Coello Coello CA (2013) Use of cooperative coevolution for solving largescale multiobjective optimization problems. IEEE Congress Evolut Comput
2. Cook W Mathematical programming computation
3. Nebro A et al (2010) A study of multiobjective metaheuristics when solving parameter scalable problems. IEEE Trans Evolut Comput

4. Coello Coello CA, Nebro AJ, Alba E, Durillo JJ, Luna F (2008) A comparative study of the effect of parameter scalability in multi-objective metaheuristics. In: Congress on evolutionary computation (CEC '2008). IEEE Service Center, Hong Kong
5. Coello Coello CA (2004) A study of the parallelization of a coevolutionary multi-objective evolutionary algorithm. Lecture notes in computer science
6. Guillermo L, Coello Coello CA (2011) Multi-objective ant colony optimization: a taxonomy and review of approaches. Series in machine perception and artificial intelligence
7. Durillo JJ (2010) Convergence speed in multi-objective metaheuristics: efficiency criteria and empirical study. Int J Num Methods Eng
8. Agarwal S, Meyarivan KDT, Pratap A (2002) A fast and elitist multiobjective genetic algorithm: NSGA–II. IEEE Trans Evolut Comput
9. Goldberg DE, Richardson J (1987) Genetic algorithms with sharing for multimodal function optimization. In: second international conference on genetic algorithms. Lawrence Erlbaum Associates
10. Deb T, Goldberg DE (1989) An investigation of niche and species, formation in genetic function optimization. In: Third international conference on genetic algorithms. San Mateo, California
11. Goldberg DE (1989) Genetic algorithms in search, optimization and machine learning. Addison-Wesley Publishing Company, Massachusetts
12. Omidvar MN, Li X, Yang Z, Yao X (2010) Cooperative co-evolution for large scale optimization, through more frequent random grouping. In: Evolutionary computation (CEC). IEEE Congress
13. Zhang Q, Li H (2009) Multiobjective optimization problems with complicated Pareto sets, MOEA/D and NSGA-II. IEEE Trans Evol Comput
14. Jin Y, Olhofer M, Sendhoff B (2016) Test problems for large-scale multiobjective and many-objective optimization Ran Cheng. IEEE Trans Cybern
15. Deb, K, Thiele L, Zitzler E (1999) Comparison of multiobjective evolutionary algorithms: empirical results, TIK-Report Number 70
16. Zhou A, Zhang Q, Jin Y (2008) RM-MEDA: a regularity model-based multiobjective estimation of distribution algorithm. IEEE Trans Evol Comput
17. Zheng J, Li M (2009) Spread assessment for evolutionary multi-objective optimization. In: Proceedings international conference on evolutionary multi-criterion optimization. Nantes, France
18. Yen GG, He Z Performance metrics ensemble for multiobjective evolutionary algorithms. IEEE Trans Evol Comput
19. Meyarivan T, Agarwal S, Pratap A, Deb K (2002) A fast and elitist multiobjective genetic algorithm: NSGA-II. IEEE Trans Evol Comput

Automatic License Plate Recognition

K. Indira, K. V. Mohan and Theegalapally Nikhilashwary

Abstract Automatic License Plate Recognition (ALPR) is of great significance in intelligent traffic monitoring systems. ALPR, a prime concern of smart surveillance, has become necessary owing to increase in vehicle theft, unlawful acts such as trespassing into restricted areas, violation of traffic rules, to ensure effective law enforcement. ALPR system for videos is generally framed in four steps: vehicle tracking, localization and extraction of license plate, character segmentation, and character recognition. Vehicle tracking plays a key role in ALPR for its direct impact on accuracy and efficiency of the system. This paper describes a novel technique for effective vehicle tracking that involves monitoring vehicle's Euclidean distance from a visible calibration line. Frames which satisfy necessary and sufficient conditions during this phase are picked up for further processing. Text extraction includes localization of the license plate based on its inherent property of greater transitions in pixel values compared to that of its surroundings. This property facilitates analyzing rows and columns in a binary image and clips them off if threshold is not met. Characters are segmented using vertical projection profile and then recognized using template matching. This method is tested on two video databases of varying resolutions and illumination levels containing a total of 100 vehicles and is found to provide an average tracking efficiency of 94.66% and average recognition accuracy of 95.41%.

Keywords License plate · Euclidean distance · Calibration line · Tracking Segmentation · Recognition

---

K. Indira · K. V. Mohan (✉) · T. Nikhilashwary (✉)
Ramaiah Institute of Technology, Bengaluru, India
e-mail: kvmohansrinivas@gmail.com

T. Nikhilashwary
e-mail: nikileshwary@gmail.com

K. Indira
e-mail: indira@msrit.edu

S. Bhattacharyya et al. (eds.), *Recent Trends in Signal and Image Processing*, Advances in Intelligent Systems and Computing 727,
https://doi.org/10.1007/978-981-10-8863-6_8

1 Introduction

Vehicle monitoring has become a tougher job for traffic controllers and toll staff due to increased number of vehicles. To ensure effective law enforcement, a better monitoring system that can help the police force to tackle situations of vehicle-related crime in much faster rate is needed. Thus, ALPR has gained utmost importance in Intelligent Surveillance Systems. Though a lot of research has been done in this area, most of them work under constrained environments making few assumptions about the context. This paper focuses on making the ALPR systems as generic as possible by reducing human intervention and environmental effects. Detection and localization of license plate from the surveillance video are difficult primarily because license plates generally occupy a small portion of the whole frame. It is a challenging task to extract license plate reliably owing to the variations in character size, font style, and presence of complex background. Characters may appear cluttered due to positioning of the camera in different orientations. Other factors that affect the system performance are varying illumination levels, poor camera quality, motion properties, and the blocking contents. Thus, building a generic ALPR system requires the algorithm to be adaptive to such variations thereby adjusting itself appropriately.

Applications of ALPR include intelligent parking management, automatic electronic toll payment, vehicle monitoring in residential areas, border monitoring and in pursuit for stolen vehicles. The process of ALPR can be performed either by sending the information to a remote computer to perform actual processing operations or at the location of camera itself where real-time operations are performed. Once ALPR is enforced on roads, the information like license plate text, date, time from the system must be obtained in roughly milliseconds. This information is useful for electronic toll collection with no human association and in intelligent traffic monitoring. The paper is arranged in the following sections: Sect. 2 gives a review of methods available to detect and recognize license plate in images and videos. In Sect. 3, proposed method of vehicle tracking in a video, text localization, text extraction, segmentation, and recognition of the registration number is discussed. Section 4 demonstrates the efficiency of the system on two types of video datasets. Conclusion and future work is presented in Sect. 5.

2 Literature Survey

Lucky and Sukadev [1] have presented full-featured vehicle detection, tracking, and license plate recognition system. Background is subtracted once the foreground is estimated using Gaussian mixture model. The license plate, extracted using block variance technique, is segmented using region-based approach. The characters are recognized using template matching algorithm. Silva et al. [2] have proposed a novel approach for ALPR using SIFT algorithm. The extracted features from key points obtained by applying SIFT algorithm on each template are stored in a database. The same process is repeated on the input image. Extracted features so obtained are compared with that of templates' extracted features to determine characters of license plate based on the regions around best matching local. The experiment was conducted on 60 plate images obtained from toll square and achieved 99.52% accuracy for location of characters in the plate and recognition accuracy of 98.57%. Kaushik and Kang-Hyun [3] have proposed segmentation of license plate region using color model. HIS color is adopted to detect the license plate regions. Classification is performed based on license plate's geometrical properties. This method detects license plate regions with different orientations and scales of the license plate. Experiments were conducted on 150 images taken from illumination, complex scenes, various environments in campus parking, and they have achieved a success rate of 94%. Harpreet and Balvinder [4] have discussed in their paper license plate detection in vehicles captured from a fixed camera using edge detection and morphological operations. Here, the videos captured from the camera are converted into frames and then frames are converted into images. License plate is then extracted using edge detection and morphological operations. The authors have reported an accuracy of 90.8% and 90% for license plate detection in a video of 240 frames per video and 200 frames per video, respectively. Nauman et al. [5] have presented Number Plate Recognition for images. Here, license plate is detected using horizontal histogram of edge image, and using morphological operation, exact license plate is detected. Binary image is obtained by choosing a threshold which is 10% of white pixels in the image. The characters are segmented using connected component analysis. They have reported 90.4% average accuracy for the license plate with Punjabi characters and 78.1% for license plate with non-Punjabi characters. An efficient real-time automatic license plate recognition for CCTV forensic application has been implemented by Saquib et al. [6]. In this paper, license plate is localized using background learning. Frames are preprocessed to improve detection efficiency as a result of enhanced plate regions. The probable text candidates are obtained using contour details of the connected components. The license plate is selected using HoG and simple nearest mean classifier. Lucas–Kanade tracker is used for vehicle tracking. The detected license plate is enhanced. Using minimum boundary rectangle, bounding boxes are drawn and individual bounding box of each character is recognized using nearest neighbor classifier. The proposed approach has achieved an overall average detection/localization accuracy of 94% and recognition accuracy of 100% for CCTV road surveillance videos. Abhay and Nidhi [7] have addressed the issue of ALPR system using RBF neural network and features of SURF. In this paper, license plate detection is performed using SURF method and plate is extracted using Hessian matrix. With this method, interested points are located in both space and sub-pixel accuracy. The outliers are detected using random sample consensus algorithm. The characters are then segmented using morphological operations, and they are then recognized using RBF neural network. The algorithm was tested with standard license plates from sources such as Google and has achieved 95% detection accuracy and 90% recognition accuracy. From related work on ALPR system, it can be observed that most of the researchers have worked on images captured from the video and only few researchers have concentrated on videos. The objective of this paper is to discuss in detail the tracking of vehicles with the license plate and extract key frame in the video with the license plate to reduce the processing onto entire video. The license plate is detected and extracted. The text is then segmented, and the registration number of the vehicles in the video is displayed.

3 Methodology

This section describes the complete process of ALPR for a surveillance video performed in four stages: vehicle tracking, license plate localization and extraction, character segmentation, and character recognition. Figure 1 shows the block diagram and the steps to be followed in ALPR system. Initially, frames in the video are preprocessed and the vehicle is tracked using line interpretation algorithm.

3.1 Preprocessing

For each frame of the video, the following preprocessing steps are performed. The RGB frame is converted to gray scale. Enhance the contrast of the image such that 1% of data is saturated at low and high intensities of original intensity level. The original frame in RGB format and the contrast enhanced and binarized frame are shown in Fig. 2a and Fig. 2b, respectively.

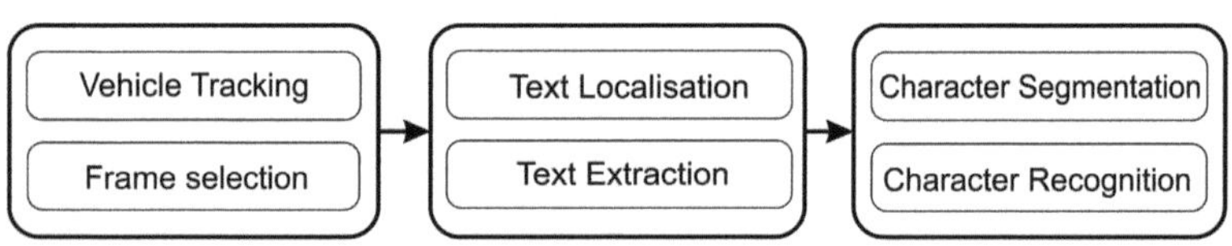

Fig. 1 Block diagram of ALPR system

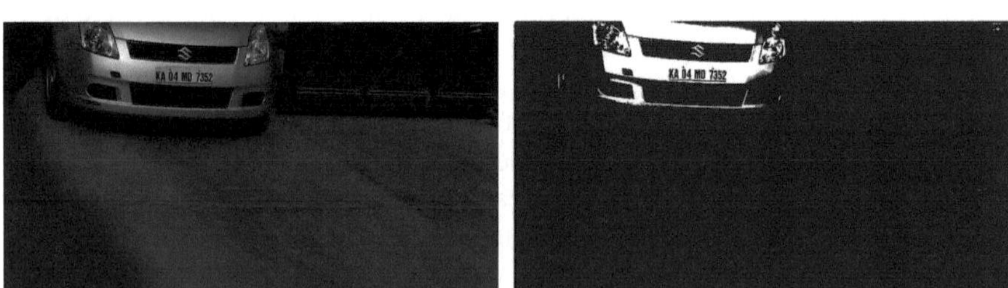

(a) Original Frame (b) Contrast enhanced and binarized frame

Fig. 2 Preprocessing frames

3.2 Line Interpretation Concept

The frame number is stored in a separate variable to keep track of the frame count. The following steps are performed on each frame of a video. A calibration line (the line in yellow) is considered with end pixel points P1 and P2 as two reference points. Figure 3 illustrates the proposed line interpretation concept. Blob analysis is used on contrast adjusted grayscale image to detect the vehicle. Using Blob analysis, bounding box is plotted around the vehicle. Centroid C is calculated using bounding box coordinates. The Euclidean distances D1 and D2 between the point C and the calibration line are measured. The Euclidean distance is defined as length of the line segment between two points. This distance is represented as a blue line. A threshold distance is to be fixed where the frame selection process must be triggered. The distance threshold is chosen such that the license plate is visible as desired, ensuring accurate results. If the ratio of distances D1 and D2 reaches the threshold, corresponding frame number and the centroid point C are recorded in an array A1. The above steps are performed repeatedly on all frames of running video and tracking of cars is performed and subsequently the frames of the car tracked are picked. There are chances of more than a single frame of the same vehicle being stored in the array. The array is processed by finding the consecutive frame number to select a key frame. If the consecutive frame number is not found, then it indicates that frame is of next vehicle. A key frame can be selected using average of total number of frames of same vehicle or by selecting the first frame itself. Only these frames are used for text extraction process, thereby reducing the computation complexity. Human intervention is avoided by constraining the aspect ratio of the detected blobs. The

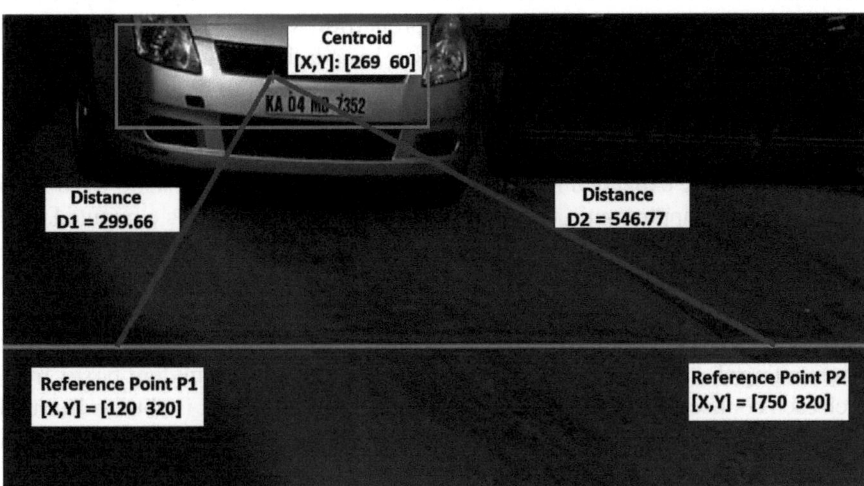

Fig. 3 Illustration of line interpretation concept

centroid point is monitored against a threshold distance rather than for the entire length of the frame to reduce unnecessary preprocessing steps during this phase.

3.3 License Plate Extraction

The vehicle tracking step provides a frame called key frame which is almost ideal for further computation. Tracking of license plate is mainly to reduce processing time in further operations. The key frame picked is first converted into gray scale and its brightness is increased to a desired level. By adjusting the contrast to a suitable range ensuring highlighted license plate, the image is binarized. The number of transitions (1–0 or 0–1) for each row is scanned. If the number of transitions is below the predefined threshold, that row is clipped off. The predefined threshold is decided based on the property that the number of transitions for text in license plate is more when compared to that of the texts surrounding in the license plate. This step is performed for all the rows in the image. Then for each column in the resultant image, the total number of 0's or 1's is counted to know the minimum and maximum column numbers where the image should be cropped columnwise. Using these min and max column indices, the resultant image which contains the number plate is cropped as shown in Fig. 4a. Accurate boundaries of license plate are determined using Maximally Stable Extreme Regions (MSER) and stroke width. The variations in the stroke width of text regions are found to be less when compared to non-text regions. Text regions with consistent color and high contrast are detected using MSER detector. Some of the geometric properties which are effective in the discrimination of non-text regions and text are extent, eccentricity solidity, aspect ratio, and Euler number. Using geometric properties, stroke width variations, and simple thresholds, non-text regions like lines or boundaries of car parts are removed. The results obtained at this stage are individual characters. To recognize them, the characters are merged to form words. One way to merge the characters into words is using neighboring text regions to form a bounding box as shown in Fig. 4b. The bounding boxes are expanded to overlap so that the text regions of same word form a chain. The overlapping ratio between all pairs of bounding box is computed. The distance between all pairs of text regions is quantified that in turn helps in finding groups of neighboring text. The text regions that are connected by a nonzero overlap ratio are found out after pairwise overlap ratios are compared. A single bounding box is formed by merging numerous neighboring bounding boxes. This is achieved by computing the maximum and minimum of the individual bounding boxes from which each connected component is made. A single bounding box is obtained around individual text lines or words by merging the overlapping bounding boxes as shown in Fig. 4c. Extracted license plate is as shown in Fig. 4d.

(a) License plate after cropping rows & columns

(b) Bounding Boxes around Text Regions

(c) Localized License Plate

(d) Extracted License plate

Fig. 4 Results of each step in extraction of license plate

3.4 License Plate Segmentation

Line segmentation is performed using horizontal projection profile, and word and character segmentation is performed using vertical projection profile with different thresholds. Vertical projection profile is applied to the extracted license plate, and the individual characters of license plate are segmented using the valleys between the characters. If the extracted number plate has two rows, horizontal profile projection (HPP) is performed prior to vertical projection profile to segment each character. The characters are segmented by fixing the threshold above the global minima (valley points) for the profile. The result of VPP on the extracted license plate is shown in Fig. 5a, and extracted characters are shown in Fig. 5b.

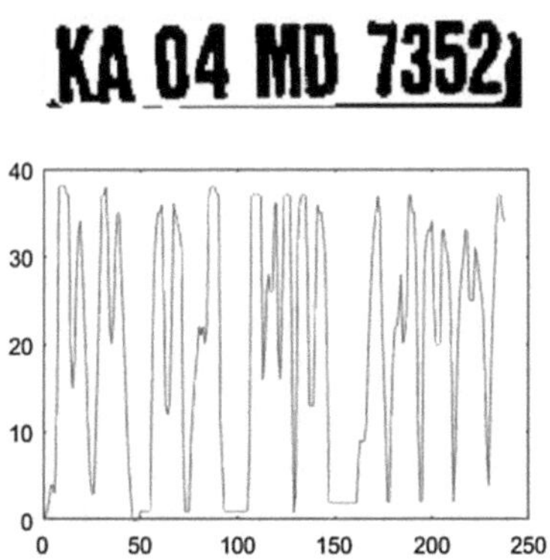

(a)Vertical Projection Profile on extracted license plate

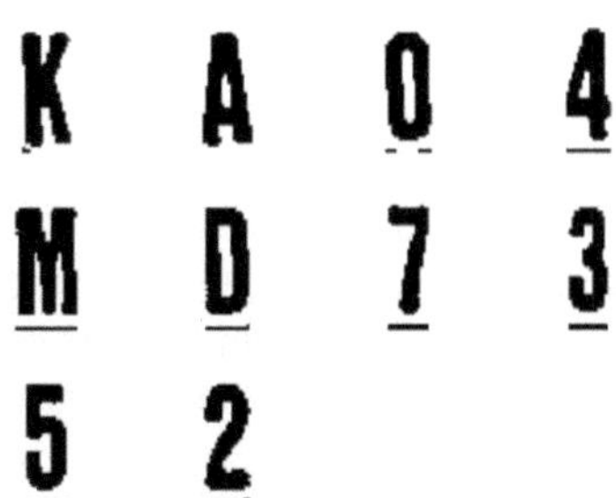

(b) Segmented characters

Fig. 5 Results of segmentation stage

3.5 License Plate Recognition

The final stage is character recognition performed using template matching. Database of standard set of all character templates is created. The correlation factor between the segmented character and template characters in the database is computed. The template character with highest correlation factor is considered as the matched character, and the character is recognized. This process is repeated for all the segmented characters, and the recognized characters are concatenated to obtain complete license plate text.

4 Experimental Results and Discussion

The proposed methodology is evaluated on two sets of database. One set is acquired during daytime with frame rate of 25 fps, with video duration of 1 h 25 min, and with a resolution of 1920 × 1080. The second set is acquired at a toll plaza during afternoon with frame rate of 25 fps, with video duration of 2 h 10 min, and with a resolution of 1920 × 1080. The number plate is tracked using line interpretation technique, and key frame is selected. The selected frame is subjected to the extraction process giving the extracted license plate. The extracted license plate is segmented, and segmented characters are recognized using template matching. The performance of the system for the two video sets is given in Table 1. From the table, it is clear that average accuracy of tracking of number plate from moving vehicles is 94.66%, average number plate detection accuracy is 96.45%, and average recognition accuracy is 95.41%. As most of the work reported in the literature is seen on frames obtained from videos, the proposed work can be compared with other works in terms of number plate detection accuracy and recognition accuracy and is given in Table 2. The success rate of proposed work is almost similar to the existing methods.
Proposed work is tested on two sets of database, one at Ramaiah of Technology, Bengaluru entrance gate, and the other at Hosakerehalli, Bengaluru toll plaza, with variations in resolution and location of camera. The two sets have 100 vehicles inclusive. Proposed method is observed to provide an average detection accuracy of 96.45% and average recognition accuracy of 95.41%. From Table 2, it is evident that the results are similar to the existing methods for the ALPR system even when generalized.

The resolution of a video dataset was reduced to 860 × 480 to check the effectiveness of the proposed method with respect to varying resolutions (Fig. 6).

Table 1 Performance summary for Dataset 1 and Dataset 2

Sl. no.	Operation	Total no. of vehicles/characters		No. of vehicles tracked/characters segmented		Success rate %	
		Dataset 1	Dataset 2	Dataset 1	Dataset 2	Dataset 1	Dataset 2
1	Vehicle tracking	25	75	23	73	92.00	97.33
2	Number plate extraction	23	73	22	71	95.65	97.26
3	Character segmentation	220	710	217	707	98.63	99.57
4	Character recognition	230	710	218	682	94.78	96.05

Table 2 Comparison of proposed work with various methods

Sl. no.	Author	Images per video	Detection accuracy %	Recognition accuracy %
1	Harpreet and Balvinder [4]	240	90.80	–
2	Kaushik and Kang-Hyun [3]	150	94.00	–
3	Naumman et al. [5]	271	92.90	96.80
4	Silva et al. [2]	60	70	88.10
5	Abhay and Nidhi [7]	–	95.00	90.00
6	Proposed work	100	96.45	95.41

(a) Original Frame with resolution 1920 x 1090

(b) Original frame with resolution 860 x 480

(c) Extracted License Plate

(d) Extracted License Plate

Fig. 6 Experimental results of proposed method on different license plates

5 Conclusion

The proposed system can track, detect, and recognize the license plate that contains font in varying style, size and color. It has been tested on two sets of database acquired from various locations at different environmental conditions with respect to illumination, background, distance between camera and the vehicle entrance or exit area. The future work includes generalizing the system to a wider extent and improving the system for simultaneous multiple vehicle tracking, followed by extraction and recognition of license plate.

6 Declaration

We hereby declare that we have obtained required permission to acquire video database from College and Tollgate respectively. We are solely responsible for any issues rise hereafter.

References

1. Lucky K, Sukadev M (2013) ITSI Trans Electric Electron Eng 1(6):2320–2327
2. Silva FA, Artero AO, Paiva MSV, Barbosa RL (2013) ALPRs–a new approach for license plate recognition using the SIFT algorithm. Signal Imag Process Int J 4(1):17–33. https://doi.org/10.5121/sipij.2014.4102
3. Kaushik D, Kang-Hyun J (2001) Segmentation of license plate region using a color model. In: Pattern recognition high performance distributed computing, pp. 181–184. IEEE Press, New York. https://doi.org/10.1109/HPDC.2001.945188
4. Harpreet KM, Balvinder S (2012) Vehicle license plate detection from video using edge detection and morphological operations. Int J Eng Res Technol 1(9)
5. Nauman S, Hassam M, Tahir HM, Umar F (2016) Automatic license plate recognition using extracted features. In: 4th international symposium on computational and business intelligence, pp 221–225
6. Saquib M, Sarfraz I, Muhammed Z, Adnan E, Edirisinghe EA (2013) Real-time Automatic License Plate Recognition for CCTV forensic applications. Int J Real Time Imag Process 8:285–295
7. Abhay C, Nidhi S (2013) Automatic license plate recognition system using SURF features and RBF neural network. Int J Comput Appl 70(27):37–41

S-shaped Binary Whale Optimization Algorithm for Feature Selection

Abdelazim G. Hussien, Aboul Ella Hassanien, Essam H. Houssein, Siddhartha Bhattacharyya and Mohamed Amin

Abstract Whale optimization algorithm is one of the recent nature-inspired optimization technique based on the behavior of bubble-net hunting strategy. In this paper, a novel binary version of whale optimization algorithm (bWOA) is proposed to select the optimal feature subset for dimensionality reduction and classifications problem. The new approach is based on a sigmoid transfer function (S-shape). By dealing with the feature selection problem, a free position of the whale must be transformed to their corresponding binary solutions. This transformation is performed by applying an S-shaped transfer function in every dimension that defines the probability of transforming the position vectors' elements from 0 to 1 and vice versa and hence force the search agents to move in a binary space. K-NN classifier is applied to ensure that the selected features are the relevant ones. A set of criteria are used to evaluate and compare the proposed bWOA-S with the native one over eleven different datasets. The results proved that the new algorithm has a significant performance in finding the optimal feature.

Keywords Whale optimization algorithm · Binary whale optimization algorithm S-shaped · Feature selection · Classification · Dimensionality reduction

A. G. Hussien (✉)
Faculty of Science, Fayoum University, Faiyum, Egypt
e-mail: aga08@fayoum.edu.eg

A. E. Hassanien
Faculty of Computers and Information, Cairo University, Giza, Egypt
e-mail: aboitcairo@gmail.com

E. H. Houssein
Faculty of Computers and Information, Minia University, Minya, Egypt
e-mail: essam.halim@mu.edu.eg

S. Bhattacharyya
Department of Computer Application, RCC Institute of Information Technology, Kolkata 700 015, India
e-mail: dr.siddhartha.bhattacharyya@gmail.com

M. Amin
Faculty of Science, Menoufia University, Shibin El Kom, Egypt

S. Bhattacharyya et al. (eds.), *Recent Trends in Signal and Image Processing*, Advances in Intelligent Systems and Computing 727,
https://doi.org/10.1007/978-981-10-8863-6_9

1 Introduction

The feature selection (FS) is considered as a process of pre-processing in the machine learning. Selecting the subset of optimal features is one of the most difficult tasks for a huge and complex dataset [1–4]. The objectives of FS are to reduce the dimensionality of data and improve performance [5, 6]. FS has been proven to effectively eliminate irrelevant and redundant features. In recent years, data have become increasingly large in both attributes and instances. FS has been successfully applied in many applications such as text categorization [7], genome projects [8], customer relationship management [9], and image retrieval [10]. Therefore, at present, FS for high-dimensional data becomes very necessary for the learning tasks of the machine.

Meta-heuristic nature-inspired algorithms are now the most widely used algorithms to solve optimization problems. There are a plenty of natural meta-heuristic algorithms that are hired with FS problem such as binary bat algorithm (BBA) in [11], binary cuckoo search algorithm (BCSA) in [12], binary flower pollination algorithm (BFPA) in [13], binary flower pollination algorithm (BFPA) [14], firefly optimization algorithm for FS [15], and sine–cosine algorithm [16]. Moreover, Emary et al. have proposed the binary ant lion and the binary gray wolf optimization [17, 18], respectively. Whale optimization algorithm is a recent nature-inspiring algorithm that mimics the humpback whale in searching for prays. In this paper, we try to propose a novel approach based on sigmoid function [19].

This paper is organized as follow: a brief overview of WOA is provided in Sect. 2. In Sect. 3 we introduce our algorithm. Results of comparing new algorithm with native one is shown in Sect. 5. In Sect. 6 we summarize the proposed work.

2 Whale Optimization Algorithm

Mirjalili et al. [20] proposed the whale optimization algorithm (WOA) inspired from the whales' behavior. Their foraging behavior is called bubble-net feeding method. However, in WOA, the current best candidate solution is set either the target prey or close to the optimum. The other will try to update their position toward the best. Mathematically, the WOA simulates the swarming behavior as follows:

$$\mathbf{D} = |\mathbf{C} \cdot \mathbf{X}^*(t) - \mathbf{X}(t)| \tag{1}$$

$$\mathbf{X(t+1)} = \mathbf{X}^*(t+1) - \mathbf{A} \cdot \mathbf{D} \tag{2}$$

where t is the current iteration, X is the position vector, X^* is the position vector coincide to the best solution found, and A and C are coefficient vectors. A and C are defined as following:

$$\mathbf{A} = 2 \cdot \mathbf{a} \cdot \mathbf{r} - \mathbf{a} \tag{3}$$

$$\mathbf{C} = 2 \cdot \mathbf{r} \tag{4}$$

where r is located randomly in the range [0, 1] and a is decreased linearly from 2 to 0 over the iterations. This algorithm as many optimization algorithms has two phases: exploration and exploitation. **The exploitation phase**: Exploitation phase divided into two processes; (1) shrinking encircling mechanism: This can be obtained by decreasing a value's according to Eq. 4. Note that a is a random value between $[-a, a]$. (2) Spiral updating position: This approach calculates the distance between the whale and the prey. A spiral equation is used to mimic the helix-shaped movement as follow:

$$\mathbf{X}(t+1) = \mathbf{D}^{l} e^{bl} \cdot \cos(2\pi l) + \mathbf{X}^{*}(t) \tag{5}$$

where l is a random number in the range $[-1, 1]$ and b is a constant. A probability of 50% to choose between either shrinking encircling mechanism or the spiral model is assumed. Consequently, the mathematical model is as follow:

$$\mathbf{X}(t+1) = \begin{cases} \mathbf{X}^{*}(t) - \mathbf{A} \cdot \mathbf{D} & if\ p < 0.5 \\ \mathbf{D}^{\mathbf{l}} \cdot e^{bl} \cdot \cos(2\pi l) + \mathbf{X}^{*}(t) & if\ p \geq 0.5 \end{cases} \tag{6}$$

where p is a random number in a uniform distribution. **The exploration phase**: In another hand side, in the exploration phase, A has used random values within $1 \prec A \prec -1$ to force the agent to move away from this location and mathematically formulated as follow:

$$\mathbf{D} = |\mathbf{C} \cdot \mathbf{X}_{rand} - \mathbf{X}| \tag{7}$$

$$\mathbf{X}(t+1) = X_{rand} - \mathbf{A} \cdot \mathbf{D} \tag{8}$$

3 S-shaped Binary Whale Optimization Algorithm

In the basic WOA, the whales move in the search space to modify their positions to any point in the space, and this is called the continuous space. As regards to the nature of the FS issues, the solutions are limited to the binary space {0, 1} values. To solve FS problem, the continuous (free position) must be transforming to their corresponding binary solutions {0, 1} and hence motivates to propose a new version of the bWOA-S. The conversion is performed by applying an S-shaped transfer function. Algorithm 1 shows the steps of the proposed S-shaped binary whale optimization. The probability of transforming the position vectors' elements from 0 to 1 and vice versa has been adopted by the transfer functions and hence force the search agents to move in a binary space (Fig. 1).

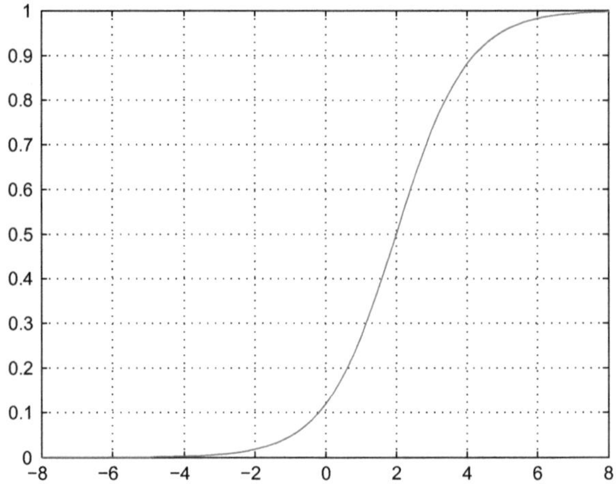

Fig. 1 S-shaped transfer function

Algorithm 1 Binary Whale Optimization S-shaped

Input: n number of whales in the population, and $MaxIter$ number of iteration.
Output: Optimal whale position.

Initialize a and n.
Find $X* = best$ search agent.
while Stopping criteria not meet to **do**
  for Whale$_i$ belong to whales **do**
    Calculate and Update a; A, C, p and l.
    if $p < 0.5$ **then**
      if $(|A| < 1)$ **then**
        Update position by Equation 2.
      else$(|A| \geq 1)$
        Choose search agent randomly (X_{rand})
        Update position by Equation 8.
      end if
    else$(p \geq 0.5)$
      Update position by Equation 5.
    end if
    Update $\mathbf{X}(t+1)$ from Equation 9
  end for
  if There exiting a search agent belong to the search space **then**
    Calculate the fitness for each agent.
    Update $X*$ if there is a better solution.
    $t = t + 1$
  end if
end while

The common S-shaped function is updating as shown in Eq. 9. Algorithm 1 shows the pseudo code of the proposed bWOA-S version.

$$y^k = \frac{1}{1 + e^{-v_i^k(t)}} \tag{9}$$

$$X_i^d = \begin{cases} 1 \; if \; rand < S(x_i^k(t+1)) \\ 0 \; otherwise \end{cases} \tag{10}$$

4 S-shaped Binary Whale Optimization Algorithm for Feature Selection

In order to solve the problem of FS, a new binary version of whale optimization algorithm bWOA-S was proposed, and hence, for feature reduction, N the number of different feature combinations would be 2^N which is a huge space of features to be searched exhaustively. So the proposed bWOA-S is applied to adaptive the search space and provided the best feature combination. The best feature combination is specifically obtained as well as the maximum classification accuracy with the minimum number of selected features. Equation 11 illustrates the fitness function used in bWOA-S to evaluate the individual whale positions.

$$Fitness = \alpha \gamma_R(D) + \beta \frac{|C - R|}{|C|} \tag{11}$$

where R is selected feature subset length, C is the total number of features, $\gamma_R(D)$ is the classification accuracy of condition attribute set R relative to decision D, and α and β are two parameters symmetric to the subset length and the classification accuracy, illustrates as $\alpha \in [0; 1]$ and $\beta = 1 - \alpha$. Therefore, this leads to the fitness function that maximizes the classification accuracy. Therefore, Eq. 11 is converted exactly into a minimization problem based on the classification error rate and selected features rather than the classification accuracy and unselected feature size. Thus, the obtained minimization problem defined as in Eq. 12.

$$Fitness = \alpha E_R(D) + \beta \frac{|R|}{|C|} \tag{12}$$

where $E_R(D)$ is the classification error rate, C is the total number of features, and R is the selected feature subset length. α and β are two parameters corresponding to the classification accuracy and feature reduction, illustrates as $\beta = 0.01$ and $\beta = 1 - \alpha$.

Eventually, a simple common K-NN classifier is applied. Therefore, K-NN is used to ensure that the selected features are the relevant ones. In the other hand, bWOA-S is adapted to maximize the feature evaluation criteria as defined in Eq. 12. Also, the individual feature is represented as a single dimension in the search space, and hence, the single feature represents the whale's position.

5 Experimental Results and Discussion

To provide a fair comparison, the experimental results are performed on different of 11 UCI benchmark datasets. Table 1 summarizes the 11 datasets from UCI machine learning that are used for the experiments and comparison, while Table 2 shows the parameter settings values. The datasets were selected to have different numbers of instances and attribute to represent various kinds of issues. In each dataset, instances are divided randomly into three different equal subsets, namely training, testing, and validation subsets in cross-validation manner. K-NN is applied in the experiment results using trial and error basis, and the best choice of K is 5, overall the different datasets. Meanwhile, every whale position produces one attribute subset through the training process. The training set is used to evaluate the K-NN classifier performance on the validation subset throughout the optimization process, and the bWOA-S is used to guide the FS process simultaneously.

Each algorithm has been performed 20 runs with random positioning of the search agents. Repeated runs of the compared algorithms were used to test the convergence

Table 1 List of data sets used in the experiments results

No.	Name	Features	Samples
1	Breastcancer	9	699
2	Tic-tac-toe	9	958
3	Zoo	16	101
4	WineEW	13	178
5	SpectEW	22	267
6	SonarEW	60	208
7	IonosphereEW	34	351
8	HeartEW	13	270
9	CongressEW	16	435
10	KrvskpEW	36	3196
11	WaveformEW	4	5000

Table 2 Parameter setting

Parameter	Value
Iterations number	70
Search agents number	8
Dimension	No. of features in the data
Search domain	$[0, 1]$
No. repetitions of runs	20
α parameter in the fitness function	0.99
β parameter in the fitness function	0.01

Table 3 Statistical mean fitness measure calculated for the compared algorithms on the different datasets using small, mixed and large initialization

No.	Small		Mixed		Large	
	WOA	bWOA-S	WOA	bWOA-S	WOA	bWOA-S
1	0.05305	0.04466	0.09977	0.04725	0.20643	0.15028
2	0.31558	0.22432	0.20914	0.20715	0.21441	0.20004
3	0.25203	0.11632	0.16824	0.139783	0.16629	0.16952
4	0.92738	0.897567	0.91507	0.915064	0.93309	0.91703
5	0.33557	0.28321	0.31313	0.307875	0.32361	0.30543
6	0.32991	0.19923	0.30304	0.278656	0.29493	0.28113
7	0.13436	0.11809	0.17146	0.155077	0.16974	0.16243
8	0.27104	0.22607	0.31744	0.297508	0.34913	0.34216
9	0.37778	0.35613	0.37771	0.371929	0.39039	0.39359
10	0.40440	0.06748	0.07364	0.06953	0.07259	0.07137
11	0.45201	0.19543	0.19458	0.19173	0.19331	0.19126

Table 4 Average classification accuracy for the compared algorithms on the different datasets using small, mixed and large initialization

No.	Small		Mixed		Large	
	WOA	bWOA-S	WOA	bWOA-S	WOA	bWOA-S
1	0.78597	0.69400	0.74091	0.62365	0.60900	0.61214
2	0.66008	0.77569	0.79885	0.80002	0.79352	0.80803
3	0.72842	0.85853	0.83688	0.82693	0.83083	0.82647
4	0.03378	0.06876	0.07879	0.07376	0.05314	0.07389
5	0.62664	0.68082	0.68478	0.66492	0.66067	0.68208
6	0.63837	0.73029	0.69856	0.71048	0.69529	0.71586
7	0.82579	0.82443	0.82449	0.83414	0.83011	0.83607
8	0.61415	0.64104	0.65244	0.64659	0.64370	0.65259
9	0.58876	0.59702	0.60087	0.58403	0.58991	0.59334
10	0.56964	0.93287	0.93437	0.93613	0.93066	0.93097
11	0.53258	0.80330	0.81143	0.81185	0.81026	0.81140

capability. In each run, eight well-known measures are recorded to investigate the performance of comparison algorithms.

Tables 3 and 4 outlines the statistical mean fitness values obtained from the two algorithms WOA and bWOA-S ,and average classification accuracy respectively in three initialization methods we can see that the proposed algorithm overcomes the native algorithm. For assessing the repeatability of results, Table 5 shows standard deviation to the obtained fitness function. There is a question here as to why we need more algorithms despite the many algorithms proposed so far. The answer to this question is in the no free lunch (NFL) theorem [21] that logically has proven that there is no optimization technique for solving all optimization problems. This

Table 5 Standard deviation fitness function for the compared algorithms on the different datasets averaged over the three initialization methods

No.	Standard deviation		Selection averaged		Average time	
	WOA	bWOA-S	WOA	bWOA-S	WOA	bWOA-S
1	0.03023	0.01226	4.72307	6.66245	0.57625	0.63875
2	0.05362	0.04834	7.13699	11.0659	0.79861	0.94444
3	0.03724	0.03441	2.89027	4.17675	0.69688	0.74765
4	0.20649	0.20237	10.05066	11.30218	0.65288	0.68750
5	0.06923	0.06874	3.36067	3.67051	0.66591	0.72159
6	0.06848	0.06318	3.64014	3.76398	0.64000	0.69604
7	0.03861	0.03683	4.21395	3.97656	0.59228	0.62610
8	0.06492	0.06904	3.18767	3.13704	0.49135	0.56250
9	0.08452	0.08923	3.85402	4.39146	0.52344	0.61797
10	0.03209	0.01394	74.21914	86.55317	0.71632	0.88924
11	0.06597	0.04137	596.25106	216.99145	0.73250	0.92344

theorem has motivated the rapidly increasing number of algorithms proposed over the last decade and is one of the motivations of this paper as well.

6 Conclusion and Future Work

In this paper, a new binary version of the basic whale optimization algorithm called bWOA-S to solve the FS problem was proposed. To convert the native version of WOA to a binary version, S-shaped transfer functions are employed. In order to investigate the performance of the proposed two binary algorithms, the experiments are applied on 11 benchmark datasets from UCI datasets and five evaluation criteria are performed. The experimental results revealed that the proposed algorithms have achieved superior results versus the native algorithm. Furthermore, the results proved that bWOA-S has been achieved the smallest number of features with better classification accuracy. For future work, the proposed algorithm introduced here will be used with more common classifiers such as SVM and ANN to verify the performance.

References

1. Chizi, B, Rokach L, Maimon O (2009) A survey of feature selection techniques. In: Encyclopedia of data warehousing and mining, 2nd edn, pp 1888–1895. IGI Global
2. Guyon I, Elisseeff A (2003) An introduction to variable and feature selection. J Mach Learn Res 1157–1182
3. Hamad, A, Houssein EH, Hassanien AE, Fahmy AA (2016) Feature extraction of epilepsy EEG using discrete wavelet transform. In: 2016 12th international computer engineering conference (ICENCO), pp 190–195. IEEE
4. Houssein, EH, Kilany M, Hassanien AE, Snasel V (2016) A two-stage feature extraction approach for ECG signals. In: International Afro-European conference for industrial advancement, pp 299–310. Springer, Cham
5. Girish C, Ferat S (2014) A survey on feature selection methods. Comput Electr Eng 40(1):16–28
6. Manoranjan D, Huan L (1997) Feature selection for classification. Intel Data Anal 1(1–4):131–156
7. Yang Y, Pedersen JO (1997) A comparative study on feature selection in text categorization. ICML, vol 97
8. Blum Avrim L, Pat L (1997) Selection of relevant features and examples in machine learning. Artif Intel 97(1):245–271
9. Kiansing N, Huan L (2000) Customer retention via data mining. Artif Intel Rev 14(6):569–590
10. Yong R, Huang Thomas S, Shih-Fu C (1999) Image retrieval: current techniques, promising directions, and open issues. J Visual Commun Image Represent 10(1):39–62
11. Nakamura Rodrigo YM et al (2012) BBA: a binary bat algorithm for feature selection. In: 2012 25th SIBGRAPI conference on graphics, patterns and images (SIBGRAPI). IEEE
12. Rodrigues D, Pereira L, Almeida T, Papa J, Souza A, Ramos C, Yang X-S (2013) BCS: a binary cuckoo search algorithm for feature selection. In: Proceedings of the IEEE international symposium on circuits and systems (ISCAS), pp 465–468. IEEE, Beijing
13. Douglas R et al (2015) Binary flower pollination algorithm and its application to feature selection. Recent Adv Swarm Intel Evol Comput, pp 85–100. Springer International Publishing
14. AbdEl-Fattah SS, Nabil E, Badr A (2016) A binary clonal flower pollination algorithm for feature selection. Pattern Recog Lett 77:21–27
15. Eid E et al (2015) Firefly optimization algorithm for feature selection. In: Proceedings of the 7th Balkan conference on informatics conference. ACM
16. Hafez, AI et al (2016) Sine cosine optimization algorithm for feature selection. In: International symposium on innovations in intelligent systems and applications (INISTA). IEEE
17. Eid E, Zawbaa HM, Hassanien AE (2016) Binary ant lion approaches for feature selection. Neurocomputing 213:54–65
18. Eid E, Zawbaa HM, Hassanien AE (2016) Binary grey wolf optimization approaches for feature selection. Neurocomputing 172:371–381
19. Seyedali M, Andrew L (2013) S-shaped versus V-shaped transfer functions for binary particle swarm optimization. Swarm Evol Comput 9:1–14
20. Seyedali M, Andrew L (2016) The whale optimization algorithm. Adv Eng Softw 95:51–67
21. Wolpert David H, Macready William G (1997) No free lunch theorems for optimization. IEEE Trans Evol Comput 1(1):67–82

Motion Artifact Reduction from Finger Photoplethysmogram Using Discrete Wavelet Transform

Anita Biswas, Monalisa Singha Roy and Rajarshi Gupta

Abstract In this work, discrete wavelet transform was used to remove the effect of motion artifact on the Photoplethysmogram (PPG) signal obtained at fingertip. Clean PPG signal and motion data (one direction) were collected from 40 healthy volunteers at 14-bit resolution using NI 6009 DAQ card, and synthetic noisy signal was generated by addition. The noisy signal was first decomposed into a specific number of levels to obtain different frequency bands. Then, soft thresholding method was used to remove the noisy components. Different wavelet functions (Daubechies, Symlet, Coiflet) and soft thresholding methods ('rigrsure,' 'heursure,' 'sqtwolog,' etc.) were used to denoise the corrupted PPG signal. A comparative study was made between all of these methods by calculating performance measures such as signal-to-noise ratio improvement, mean square error, and percentage noise retention. The mother wavelet 'Db6' and 'rigrsure' soft thresholding method showed the best result.

Keywords Photoplethysmogram · Motion artifact · Discrete wavelet transform
Thresholding · Denoising

1 Introduction

During the last two decades, the Photoplethysmography (PPG) technique has become an emerging area of research for its wide application in clinical physiological monitoring [1]. PPG is a simple, noninvasive optoelectronic method to detect the blood volume changes on the peripheral body parts. The PPG sensor detects change in light intensity due to blood volume changes as a result of cardiac systole and diastole

A. Biswas (✉) · M. Singha Roy · R. Gupta
Department of Applied Physics, University of Calcutta, Kolkata, India
e-mail: anitabiswas.biswas8@gmail.com

M. Singha Roy
e-mail: monalisa_sr03@yahoo.co.in

R. Gupta
e-mail: rgaphy@caluniv.ac.in

S. Bhattacharyya et al. (eds.), *Recent Trends in Signal and Image Processing*,
Advances in Intelligent Systems and Computing 727,
https://doi.org/10.1007/978-981-10-8863-6_10

[2, 3] using near red and infrared wavelength light source and detector. It is found that PPG waveform can provide a lot of valuable clinical information such as blood pressure, heart rate, respiratory rate, and cardiac output [4–6].

However, for ubiquitous and real-time measurement system, PPG signals are easily prone to motion artifacts (MA) which could lead to inaccurate interpretation of the PPG waveform. The common sources of errors are ambient light at the photodetector, poor contact of the PPG sensor and the skin, patient's movement, respiration, etc. [7, 8]. The motion artifacts which mix the raw PPG signal that resides in the same frequency range with the actual PPG data (0–2 Hz). In particular, the motion artifact reduction is the most challenging issue. Over the last few years, many researchers focused on the area of motion artifacts removal technique from PPG signal.

The adaptive filtering technique used in real-time application is a well-known method for motion artifacts removal. But it requires an additional sensor to provide reference signal which is the major drawback of this method [9, 10]. AS-LMS adaptive filter eliminates this drawback by generating the reference signal internally from the MA-corrupted PPG itself. Principal Component Analysis (PCA) is a popular signal processing tool that can be effectively used for noise elimination, signal separation, and feature extraction. By selecting suitable number of principal components, the dimension of the actual data can be reduced but also retaining maximum possible information of the actual data [11]. In [12], the frequency characteristics of the PPG signal are analyzed to characterize the clean PPG and motion artifact. It is observed that the raw PPG has fundamental, second and third harmonic structure caused by arterial blood flow and component caused by motion artifact. The raw signal is divided into different groups depending on the relative location of these frequency components. The motion artifact is composed of single frequency component. Another frequency domain analysis method named Empirical Mode Decomposition (EMD), that empirically extracts the oscillatory behavior from the signal and decompose a time series data into a number of intrinsic mode function (IMFs). In [13], Hilbert Hung Transform is used to generate instantaneous frequency for each IMF. The PPG signal is divided into six scales of different frequency bands. According to mean frequency, the scales 1–3 match with clean PPG signal and its harmonics and scales 4–6 correlate with the motion artifacts. An improved algorithm is presented in [14] which combine PCA with the EMD method to generate a smaller group of orthogonal variables. A fewer number of principal components are able to represent the clinical information of the signal, and this number is always less than the number of IMFs. In all of these frequency domain methods, the Fast Fourier Transform (FFT) is used to convert the signal from time domain to frequency domain. Wavelet transform is a powerful tool that has been widely used in ECG signal denoising. However, its application in PPG preprocessing is comparatively fewer [15, 16]. In [16], application of wavelet transform for MA reduction is described. However, the authors only evaluated the systolic peak amplitude variations and spectrum analysis and no detailed study on error parameters was done.

In this work, we used synthetic noisy PPG by imposing three different noise levels and decompose the signal using DWT. The performance of different mother wavelets and different thresholding rules on signal reconstruction quality was also investigated. Additionally, a detailed study on error parameters was done to evaluate the effectiveness of the proposed method.

2 Methodology

2.1 *Data Collection Protocol*

The PPG signal was recorded from forty male and female volunteers in the age group 20–45 years. Prior to data collection, they were requested to sit in a relaxed condition. A transmission type PPG sensor was placed on the left hand index finger, kept in stationary condition, while an accelerometer was attached on the right hand index finger. For our application, we considered only horizontal movement of the hand (X-axis). The right hand was moved from elbow at three predefined levels of motion (level, i.e., 10, 15, and 20 cm), each non-periodic. For every subject, data collection was done for 30 s. at 60 Hz using NI6009 DAQ device. The motion data was added with the clean PPG signal in different proportions to generate synthetic noisy signal. The PPG data was stored in calibrated voltage format against sampling time in <*.txt> formatted file.

2.2 *Wavelet Denoising Approach*

Because the PPG signal is non-stationary in nature, the discrete wavelet transform (DWT) is an appropriate signal processing tool for analyzing the time–amplitude information from a signal. It analyzes the signal at different resolutions by decomposing it into various successive frequency bands. The signal is passed through a series of high-pass filter and low-pass filter.

Since in DWT the signal is a discrete time function, it can be represented as x[n], where n is an integer. The signal is passed through a half band digital low-pass filter with impulse response h[n]. The convolution operation in discrete time is defined as follows:

$$x[n] * h[n] = \sum_{k=-\infty}^{\infty} x[k] \cdot h[n-k] \tag{1}$$

The half band low-pass filter removes all the frequencies that are above half of the highest frequency in the signal.

The denoising algorithm is given below:

a. Decomposition—The input signal was decomposed up to level 5 (to get a frequency level around 1.7 Hz, that of PPG signal) by choosing appropriate wavelet

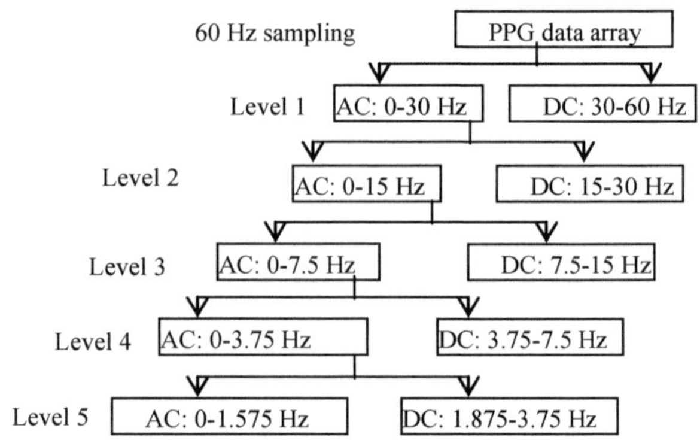

Fig. 1 Wavelet decomposition structure of the PPG data

function using Daubechies6 (Db6) as the basis wavelet. The decomposition structure is shown in Fig. 1.

b. Thresholding—In 'soft' thresholding, the coefficients below a threshold value are set to zero, while the magnitude of others is truncated by the same value. Mathematically, it is expressed as

$$\begin{aligned}\hat{d}_{j,k} &= sign(d_{j,k})|d_{j,k} - t|,\ if\ |d_{j,k}| > t \\ &= 0,\ otherwise\end{aligned} \tag{2}$$

where $d_{j,k}$ represents the kth wavelet coefficient in jth level, $\hat{d}_{j,k}$ is the thresholded value, and t is the threshold. There are various soft thresholding rules, like fixed threshold ('sqtwolog'), 'rigrsure' based on Stein's unbiased risk estimate, 'heursure' a combination of first two rules, and 'minimax' based on mean square error [17].
The soft thresholding rules were selected for our application, and it is applied in each level (A5, D5, …, D1) for selective truncation of the wavelet coefficients. The thresholded coefficients at different levels are shown in Fig. 2.

c. Reconstruction—After addition of the selected coefficients at each level, the reconstruction was done by the inverse discrete wavelet transform (IDWT).

3 Results and Discussions

For evaluating the denoising performance, the objective assessment was done by computing the following metrics which are common for any denoising operation.

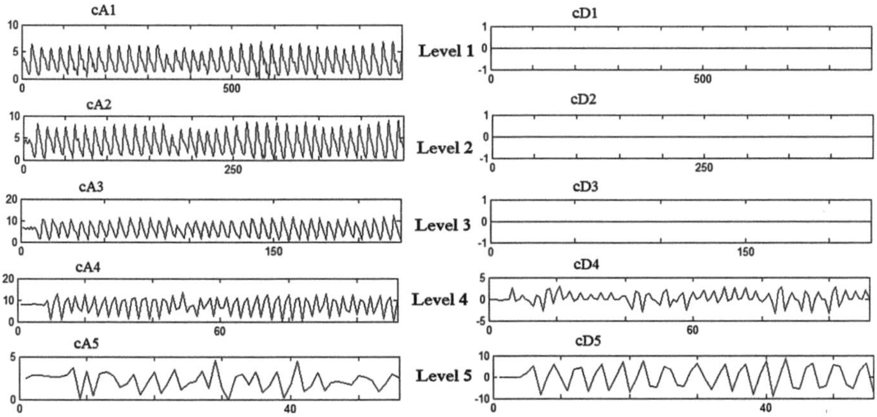

Fig. 2 Soft thresholding applied at different levels using Db6

$$
\begin{aligned}
&MSE\ (Mean\ Square\ Error) = \frac{1}{N}\sum_i [x(i) - \tilde{x}(i)]^2 \\
&MAE\ (Maximum\ Absolute\ Error) = \max[x(i) - \tilde{x}(i)] \\
&PNR\ (Percentage\ Noise\ retention) = \frac{P_{ds} - P_{cs}}{P_{cs}} \times 100 \\
&SNR_i\ (Signal\ to\ noise\ ratio\ improvement) = 10 \times \log \frac{\sum_i \frac{x^2(i)}{n^2(i)}}{\sum_i \frac{x^2(i)}{[x(i)-\tilde{x}(i)]^2}}
\end{aligned} \tag{1}
$$

where x = clean signal, $\tilde{x}$ = reconstructed signal, P_{cs} = power of clean signal, N = number of samples, P_{ds} =power of reconstructed signal, n = noisy signal, and P = $10 \times \log\sum|x(i)|^2$. Among these, MSE provides the global error figure, since errors can be positive or negative. PNR provides the residual noise that is retained in the preprocessed signal. SNR_i actually denotes the ratio of input and output SNR (to the preprocessing algorithm), where $x(i)-\tilde{x}(i)$ is the residual noise. MAE is the maximum absolute sample-to-sample error after reconstruction and local residual noise. Table 1 shows these parameters for six arbitrarily chosen healthy volunteers' data for three levels of motion using 'rigrsure' thresholding rule. Among the parameters, we observed that the MAE, averaged over the six volunteers, remains almost unchanged, while MSE and PNR increase slightly with enhanced noise, as expected. Computed over all volunteers' data, for level 1 noise, the average MSE, PNR, MAE, and SNR are 0.104, 2.05, 0.89 mV, and 14.12, respectively. When the noise level increases to level 2, these values are 0.143, 2.44, 0.983 mV, and 11.45, respectively. For highest level 3, these values are 0.230, 2.855, 0.910 mV, and 12.54, respectively. In summary, with noise-level increase, the change in MSE, PNR, MAE, and SNR are 37%, 19%, 10%, and 18%, respectively, from level 1 to 2 and 60%, 16%, 7%, and 9.51%, respectively, for level 2 to level 3. Among these, the MAE is least changed, which means the local distortion could be kept lower in the denoising process.

Table 1 Denoising parameters for db6 mother wavelet function

Volunteers' detail (age, M/F, weight in Kg, height in cm)	Level of noise	MSE	PNR	MAE (mV)	SNR
#1 (27, M, 59, 183)	1st	0.09	1.16	0.92	11.26
	2nd	0.19	2.76	1.12	8.29
	3rd	0.21	3.09	1.01	5.08
#2 (43, M, 72, 173)	1st	0.18	2.85	1.05	7.84
	2nd	0.20	2.98	1.16	7.26
	3rd	0.16	2.24	1.17	7.43
#3 (29, F, 55, 167)	1st	0.23	3.40	1.02	5.06
	2nd	0.30	3.77	1.47	4.33
	3rd	0.43	6.89	0	2.43
#4 (24, M, 65, 167)	1st	0.06	1.83	0.65	15.92
	2nd	0.06	1.71	0.38	14.27
	3rd	0.11	2.77	0.77	10.39
#5 (24, M, 68, 170)	1st	0.14	2.58	1.01	33.07
	2nd	0.10	3.45	0.60	20.54
	3rd	0.11	0.33	1.07	19.33
#6 (24, F, 60, 157)	1st	0.07	0.79	0.97	11.27
	2nd	0.09	0.27	1.19	15.03
	3rd	0.18	1.68	1.44	36.59

The denoising performance is also pictorially represented using Fig. 3 for volunteer 1. The upper panel shows the original and reconstructed signal superimposed, with the residual, i.e., the sample-to-sample error in lowest panel.

Table 2 shows the denoising parameter values for different noise levels using various thresholding rules for volunteer #1 using Db6 wavelet. Here, we studied volunteer 1 data for different levels of noise. It is clearly observable from Table 1 and 2 that MSE value is minimum for 'rigrsure' thresholding rule, while 'heursure' provides the least performance. For local distortion assessment, the fixed thresholding provides worst performance, but its SNR improvement is very good (Fig. 4).

In the next phase, four clinical features: systolic peak height, diastolic peak height, peak-to-peak time, and crest time of the reconstructed data were compared with the original signal to assess the percentage deviation for five volunteers as in Table 1. The used noise levels are level 1 at a proportion of 100 and 200%. The deviations are shown in Fig. 5. It is observed that for volunteer #1, there is no variation of peak-to-peak time event for 200% of noise, while the other features slightly increase. For volunteer #2, the crest time changes by 25% in denoised data. As a general observation, using level 1 noise at 200% proportion, most of the features could be retained within 15% of their original values.

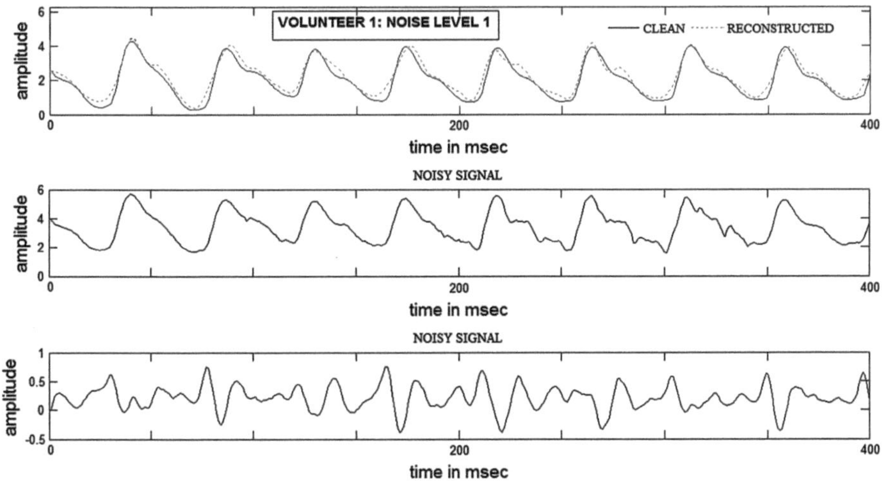

Fig. 3 Subjective assessment of denoising performance

Table 2 Denoising parameters for different thresholding rules (Volunteer #1, Db6)

Level of noise	Threshold rule	MSE	PNR	MAE (mV)	SNR_i
1st	sqtwolog	0.77	2.78	2.21	2.21
	heursure	0.92	2.09	0.92	0.92
	minimax	0.39	1.53	1.53	1.53
2nd	sqtwolog	0.45	0.62	1.77	1.77
	heursure	0.19	2.33	1.12	1.12
	minimax	0.24	0.86	1.19	1.19
3rd	sqtwolog	0.48	1.07	1.50	1.50
	heursure	0.21	2.60	1.02	1.02
	minimax	0.24	1.02	1.21	1.21

Table 3 shows the different denoising parameters for various mother wavelet functions for volunteer #1 and noise level 1. It is observed from Tables 1 and 3 that the denoising parameters show better result for Db6 mother wavelet. In terms of MSE, the Symlets 2, 3, and 4 provide the worst performance. The denoising performance is also graphically shown in Fig. 5. In the published literature using DWT-based PPG denoising [15, 16], no quantitative performance is provided. Only [9] reports SNR of 1.732 dB with horizontal motion of PPG sensor and [11] reports SNR improvement from 0.078 to 0.318 dB. The proposed method achieved an average SNR_i of 7.40 dB.

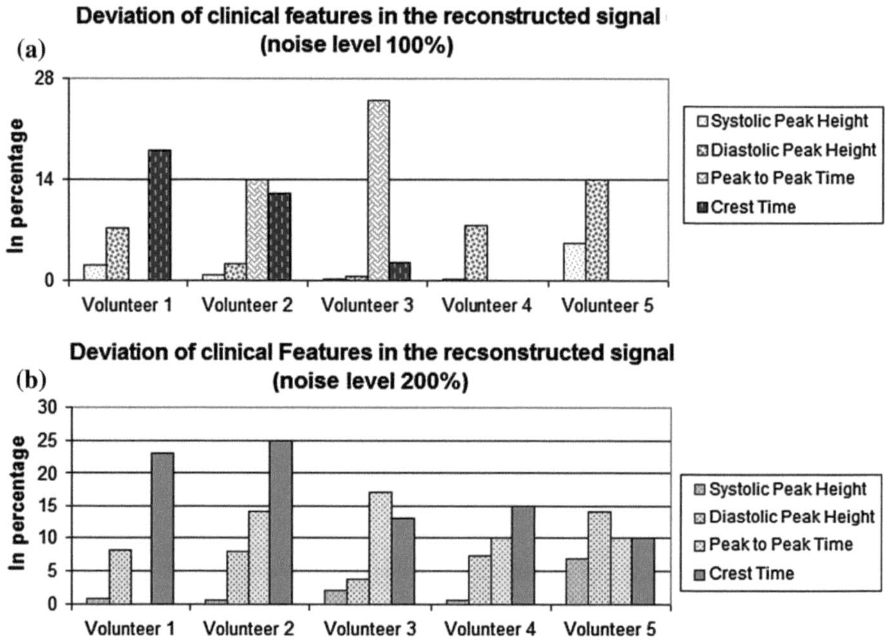

Fig. 4 Deviation of clinical signatures in the reconstructed signal for five volunteers

Table 3 Denoising parameters for different mother wavelet functions (volunteer #1)

Mother wavelets	MSE	PNR	MAE (mV)	SNR_i
Db5	1.27	4.62	2.54	0.25
sym1	3.50	2.10	5.40	7.72
sym2	4.42	0.09	4.59	25.12
sym3	6.10	9.57	4.89	7.58
sym4	2.05	4.76	6.15	11.16
sym5	5.01	4.69	4.60	0.52
coif1	1.17	2.84	3.11	8.82
coif2	2.19	6.50	5.18	11.16
coif3	1.76	4.24	3.43	4.02
coif4	2.37	5.75	4.15	1.98
coif5	2.38	3.41	4.13	3.09

4 Conclusion

In this paper, we present the application of wavelet transform for reduction of motion artifact from PPG signal. Among all wavelets used in the study, the Daubechies wavelet provided best result using 'rigrsure' soft thresholding rule. Using three

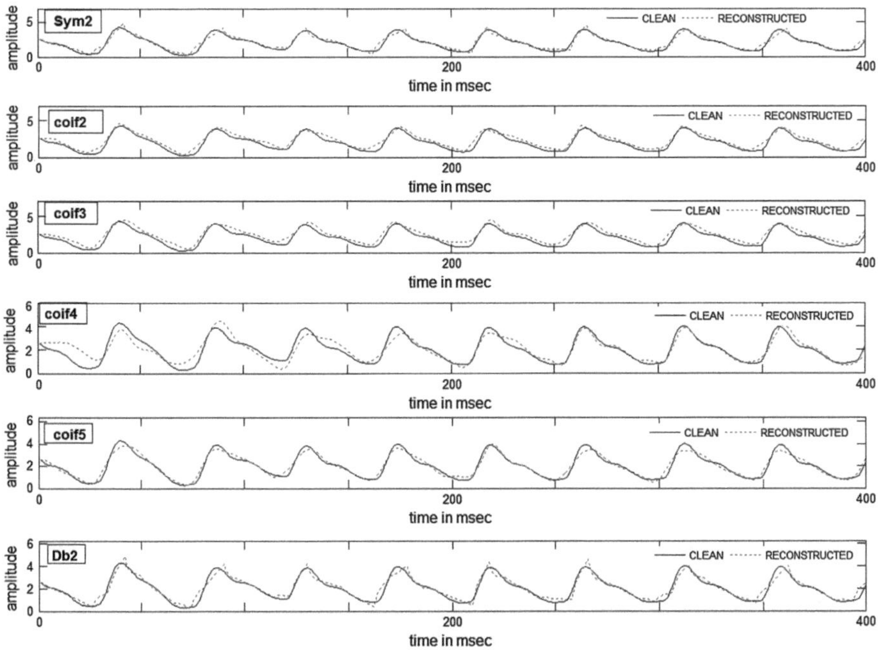

Fig. 5 Denoising performance of different mother wavelet functions

different intensity levels of noise, the reconstructed waveform could retain the most of clinical signatures within 15% of their respective values. We have also made a comparative study between different denoising parameters by using other mother wavelet functions.

References

1. Allen JJ (2007) Photoplethysmography and its application in clinical physiological measurement. Physiol Meas 28(3):1–39
2. Kyriacou PA (2006) Pulse oximetry in the oesophagus. Physiol Meas 27(1):1–35
3. Hertzman AB (1938) The blood supply of various skin areas as estimated by the photoelectric plethysmograph. Am J Physiol 124:328–340
4. Hee X, Goubran RA, Liu XP (2014) Secondary peak detection of PPG signal for continuous cuffless arterial blood pressure measurement. IEEE Trans Instrum Meas 63(6):1431–1439
5. Gil E, Orini M, Bailon R (2010) Photoplethysmography pulse rate variability as a surrogate measurement of heart rate variability during non-stationary conditions. Physiol Meas 31(9):1271–1290
6. Nakajima K, Tamura T, Miike H (1996) Monitoring of heart and respiratory rates by photoplethysmography using a digital filtering technique. Med Eng Phys 18(5):365–372 Elsevier
7. Sweeney KT, Ward TE, MacLoone SF (2012) Artifact removal in physiological signals-practices and possibilities. IEEE Trans Inf Technol Biomed 16(3):488–500

8. Han H, Kim MJ, Kim J (2007) Development of real time motion artifact reduction algorithm for a wearable photoplethysmography. In: IEEE EMBS conference France, pp 1538–1541
9. Raguram M, Madhav KV, Krishna EH, Reddy KN, Reddy KA (2011) On the performance of time varying step–size least mean squares (TVS-LMS) adaptive filter for MA reduction from PPG signals. In: ICCSP Conference, pp 431–435
10. Raguram M, Madhav KV, Krishna EH (2012) A novel approach for motion artifact reduction in PPG signals based on AS-LMS adaptive filter. IEEE Trans Instrum Meas 61(5):1445–1457
11. Castells F, Laguna P, Sornmo L (2007) Principal component analysis in ECG signal processing. Eurasip J Adv Signal Process 2007(1):98–98
12. Cho JM, Sung YK, Shin KW (2012) A preliminary study on photoplethysmogram (PPG) signal analysis for reduction of motion artifact in frequency domain. In: IEEE EMBS conference on IECBES, pp 28–33
13. Wang Q, Yang P (2010) Artifacts reduction based on empirical mode decomposition (EMD) in photoplethysmography for pulse rate detection. In: IEEE EMBC conference, pp 959–962
14. Pinheiro E, Postolache O, Girao P (2012) Empirical mode decomposition and principal component analysis implementation in processing non-invasive cardiovascular signals. Measurement 45(2):175–181. Elsevier
15. Lee CM, Zhang YT (2003) Reduction of motion artifacts from photoplethysmographic recordings using a wavelet denoising approach. In: IEEE EMBS Asian-Pacific conference on biomedical engineering, pp 194–195
16. Raghuram M, Madhav KV, Krishna EH (2010) On the performance of wavelets in reducing motion artifacts from photoplethysmographic signals. In: IEEE international conference on ICBBE, China, pp 121–128
17. Wavelet toolbox user guide MATLAB®. www.mathworks.com

Analysis of Picosecond Pulse for ATR Using Ultra-Wideband RADAR

Kaustubh Bhattacharyya, Rima Deka and Sunanadan Baruah

Abstract Automatic target recognition (ATR) permits precision strikes against specific targets with minimized risk and increased efficiency without adverse effects on other entities. Hence, it has become increasingly important for modern defense strategy. With the advances in computer technology, real-time target classification and recognition becomes an important and essential feature of radar systems, specifically for military purposes. High-frequency signal such as THz signal is a recent demand for new surveillance systems, which has led to the development of various prototypes capable of detecting concealed weapons. Radar operated in this frequency provides the possibility of higher precision signal processing due to the wider bandwidth. In this paper, analysis of the spectrum of THz pulses used for detecting radar target through simulation is presented.

Keywords Radar · ATR · LFM · Target · Terahertz · Signal processing

1 Introduction

Due to the availability of the frequency band, THz communication is receiving the attention of researchers today [1, 2]. For future wireless systems with data rates of 10 Gbps or higher, the frequency band of 275–3000 GHz is drawing attention for possible use in future communication systems. Till date, this particular frequency band has not been allocated for any specific applications. Recent development of wireless communication and VLSI technology in THz band frequency has opened up a wide application areas basically in the fields of security and surveillance [3, 4].

K. Bhattacharyya (✉) · R. Deka · S. Baruah
Assam Don Bosco University, Azara, Guwahati, India
e-mail: kaustubh.bhattacharyya@dbuniversity.ac.in

R. Deka
e-mail: rimadekas@gmail.com

S. Baruah
e-mail: sunandan.baruah@dbuniversity.ac.in

S. Bhattacharyya et al. (eds.), *Recent Trends in Signal and Image Processing*,
Advances in Intelligent Systems and Computing 727,
https://doi.org/10.1007/978-981-10-8863-6_11

Terahertz (THz) radar provides the possibility of higher precision in detecting target [5] as it can penetrate certain material clothes and reflected back by the skin [6]. Due to increase in criminal and terrorist activities, ATR with THz frequency radar system becomes an active research area, as this can serve as a reliable technique for detection of concealed weapons [7]. Hence, real-time target classification and recognition becomes an important and essential feature (function) of radar systems [8].

Appropriate feature extraction from radar returns is a key technique in ATR. Feature extraction mainly depends on the proper processing of the radar signals in time as well as in frequency domain [9]. This work describes the process of picosecond pulse to formulate a series of steps for processing, so that these can be used for UWB RADAR to discriminate among targets. Experimental results prove its effectiveness as necessary requirement for ATR with UWB RADAR. The proposed work also includes the simulation and analysis of radar transmitter and receiver. Limited works are reported in the literature [10–12] related to the processing of high-frequency (THz) pulse for ATR with UWB RADAR; hence, the present work is a novel attempt in this regard.

The rest of the paper is organized as follows: Sect. 2 describes briefly about the analysis of simulated picosecond pulse, while Sect. 3 furnishes detail about the simulation of transmitter, target environment, and receiver model followed by the experimental results in Sect. 4. Section 5 gives a conclusion of the work described as part of the proposed work.

2 Simulation and Analysis of Picosecond Pulse

LFM chirp has a spectrum that is good approximation to a rectangle function. This is only true if the time–bandwidth product $\beta\tau$ (β = swept instantaneous bandwidth, τ = pulse length) is reasonably high [13]. The swept bandwidths considered here are β_i = 300 GHz, 1 THz, and 5 THz with τ = 100 ps, giving time–bandwidth (TB) products of 30, 100, and 500. In practice, a TB product of 100 or better is usually considered large enough to model the LFM spectrum as a rectangle. An LFM with a TB product of 500 was compared with a simple pulse of the same duration and sampling rate. Both waveforms have duration of 100 ps; the LFM chirp has a bandwidth of β = 5 THz to give a TB product of 500. It was started by constructing the matched filter for each pulse and computing the response of this filter to the echo from a single point target. The difference in shapes is dramatic, but can be noted that the peak value of both outputs is exactly at 5000 as shown in Fig. 1.

The LFM chirp is called a "pulse compression" [14] waveform because at the output of the matched filter, most of its energy is compressed into a much narrower spike than is the case for the simple pulse of the same length. Consider two scatterers 10 ps apart in received echo time corresponding to (c/2) $\times$ 10 ps = 1.5 mm of range separation. It was found that the simple pulse completely fails to resolve the two targets, while the LFM chirp resolves them easily, indicating that two distinct and

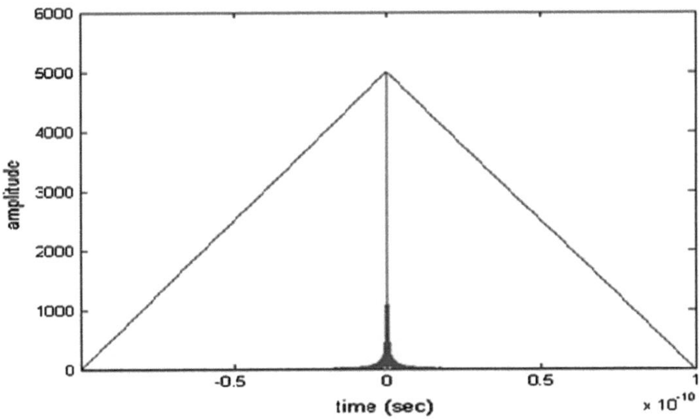

Fig. 1 Comparison of respective matched filter outputs to chirp and simple pulses of the same length

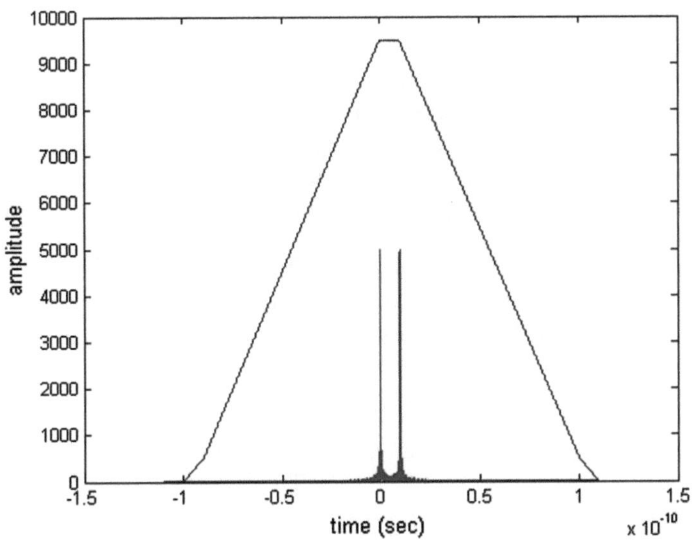

Fig. 2 Output of matched filter for the same two pulses and two targets placed 1.5 mm (10 ps) apart in range

easily separable peaks are visible at the matched filter output with the correct spacing of 10 ps as shown in Fig. 2.

The rectangle pulse has a number of side-lobes which will affect the target recognition creating some kind of harmonics; hence, the SNR will be affected. The windowing technique was used for suppression of the side-lobe. The window was applied to the filter response in the time domain in order to reduce the filter side-lobe in the frequency domain. One way to do this would be to perform conventional matched

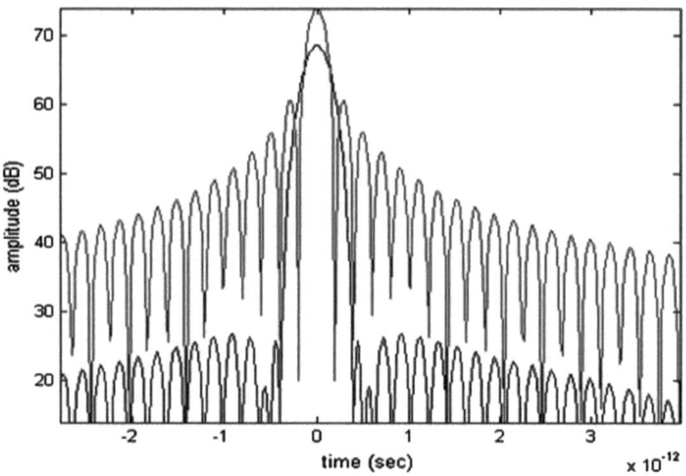

Fig. 3 Zoomed version of the output of frequency-domain matched filter, with and without Hamming weighting

filtering in the time domain; compute the DFT of the matched filter output; apply the window to that spectrum; and then take the inverse DFT to obtain the reduced-side-lobe response. However, to reduce the required computation, the matched filter operation can be combined with the windowing by performing the matched filtering in the frequency domain.

Since the window modifies the matched filter frequency response, there is an unavoidable loss in the absolute peak of the windowed filter response. Hence, it is no longer exactly matched to the transmitted waveform. In Fig. 3, a loss of 7 dB was noted as the peak reduced from 75 to 68 dB. The predicted loss can be calculated using Eq. (1):

$$LPG = \frac{1}{K^2} \mid \sum_{k=0}^{K-1} w[k] \mid^2 \tag{1}$$

The first side-lobe for the unwindowed case is 13 dB below the peak, and for the windowed case, it is 43 dB below the peak. The peak side-lobe (unlike the first side-lobe in the weighted case) goes down by 40 dB with weighting. Thus, the use of the Hamming weighting shows improvement in peak side-lobe suppression by 26.5 dB (40–13.5).

Unweighted time-domain matched filtering was applied to a signal consisting of the same up-chirp used above, but with a Doppler shift corresponding to 10% of the swept bandwidth (thus, 0.5 THz) imposed on the returned echo. This is a large Doppler shift with reference to the 5 THz bandwidth at common radar frequencies, but is convenient for clarity.

As a result of the range-Doppler coupling phenomenon, it is expected that the Doppler shift to shift the peak of the matched filter output response away from its correct location. The predicted shift in the peak location is

$$\Delta tpeak = -(Fd \times \tau) \div \beta = (-\beta \div 10)(\tau \div \beta) = -\frac{\tau}{10} = -10\,\text{ps} \quad (2)$$

where the particular value of $\tau = 100$ ps was used in the last step of Eq. (2). It is thus expected that the matched filter output peak would occur 10 ps early. The leftmost peak in Fig. 4 is the response of the matched filter for the up-chirp waveform and the specified Doppler shift. The peak occurs at -10 ps as predicted. The results plotted in Fig. 4 suggest a way to compensate for the erroneous range measurements (peak locations) that are caused by uncompensated Doppler shifts on the received signal. At least for isolated targets, we could transmit an up-chirp on one pulse, and the corresponding down-chirp on the next pulse. The true range location would then be the average of the two measured peak locations, since they will be displaced from the correct location by equal amounts but in opposite directions. The second (rightmost) peak in Fig. 4 is the result when a down-chirp with the same swept bandwidth and duration as the original up-chirp was used; the direction of the frequency sweep was simply reversed. The same positive Doppler shift is applied to the waveform, and the peak then occurs at +10 ps.

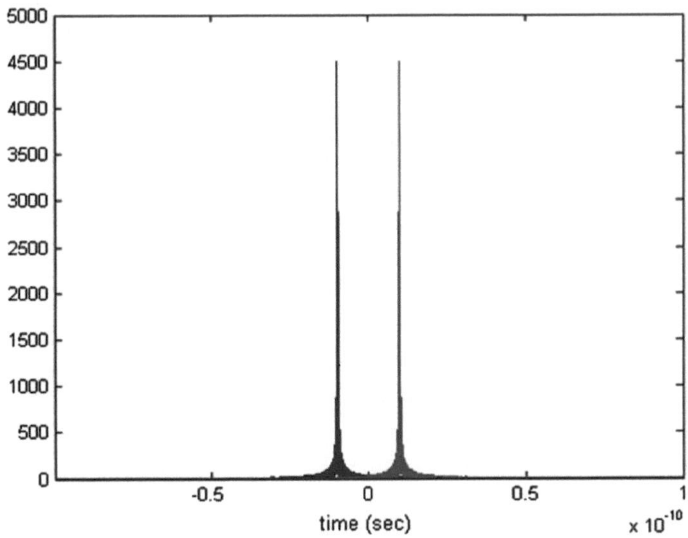

Fig. 4 Output of matched filters to the same Doppler shift, but measured with complementary up- and down-chirps

3 Simulation of Transmitter, Receiver, and Target Environment

- ***Radar Transmitter***: The peak transmitted power of a transmitter is the most critical parameter. Many factors including maximum unambiguous range, the SNR at receiver, and the pulse width of the transmitted waveform govern the required peak power. The design goal of probability of detection (P_d), probability of false alarm (P_{fa}), and the detection mechanism adopted at the receiver [14] determines required SNR at the receiver. The receiver operating characteristics (ROC) curve [14] as shown in Fig. 5 best represents the relation between P_d, P_{fa}, and SNR.
- ***Radar Receiver***: It was assumed that the thermal noise is the lone noise present at the radar receiver to remove clutter in this simulation. There is an interrelation between the thermal noise power and the bandwidth of the receiver. The receiver's noise bandwidth is set at the same value as the bandwidth of the waveform, conforming to real systems. A 20 dB gain and a 0 dB noise are assumed for carrying out the simulation. Since a monostatic radar is modeled, the receiver can be turned on only when the transmitter is off [15].
- ***Radar Target***: The radar cross section is a measure of size and ability of a target to reflect its energy. Radar cross section σ in m^2 is a measure of detectability of an object [15]. A higher value of radar cross section assures easier detection of an object [16, 17]. The reflected power from the target depends on different factors, such as [18]:

  – material and absolute size of the target;
  – relative size of the target with respect to the wavelength of the transmitted signal;

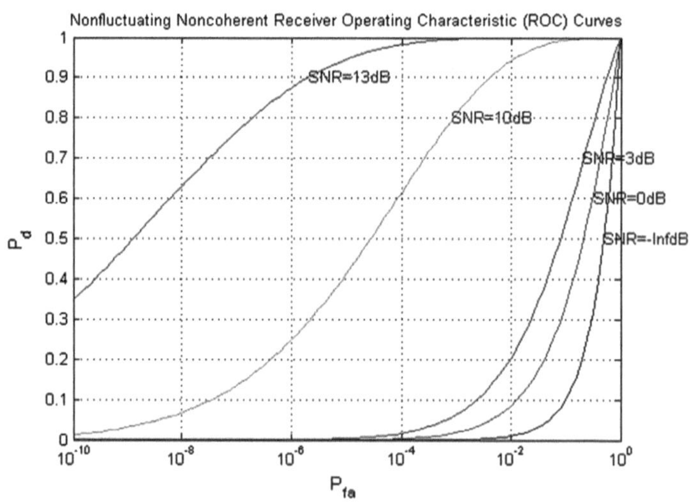

Fig. 5 ROC curve without numerical pulse integration technique

Table 1 Stationary target status

Target	RCS (in m^2)	Position in space (x:y:z)
Target1	5.05	[116:66:0:0]

Table 2 Polarized target status

Target	Mode	Scattering matrix source	Scattering matrix	Operating frequency (Hz)
Target2	Bistatic	–	[1 0:0 1]	$3 * 10^{12}$
Target3	Bistatic	Input port	–	$3 * 10^{12}$
Target4	Monostatic	Input port	–	$3 * 10^{12}$
Target5	Monostatic	Input port	–	$3 * 10^{12}$

- the incident and reflected angle;
- the polarization of transmitted and the received signal with respect to the orientation of the target.

For the target detection system, we have defined five targets of all different characteristics. One of the targets is stationary, and other targets are polarized target with different properties as given in Tables 1 and 2.

The reflected signal from a target with a gain factor of G for an incoming signal X is given by:

$$y = \sqrt{G \cdot X} \tag{3}$$

For polarized waves, the scattering equation is more complicated. The single scalar signal, X, is replaced by a vector signal (E_H, E_V) with horizontal and vertical components. A scattering matrix, S, replaces the scalar cross section, σ. Through the scattering matrix, the incident horizontal and vertical polarized signals are converted into the reflected horizontal and vertical polarized signals:

$$\begin{bmatrix} E_H \\ E_V \end{bmatrix} = \sqrt{\frac{4\pi}{\lambda^2}} \begin{bmatrix} S_{HH} & S_{VH} \\ S_{HV} & S_{VV} \end{bmatrix} \begin{bmatrix} E_H \\ E_V \end{bmatrix} = \sqrt{\frac{4\pi}{\lambda^2}} S \begin{bmatrix} E_H \\ EV \end{bmatrix} \tag{4}$$

4 Results and Discussion

The spectral analysis of the reflected signal was carried out, and from Figs. 6 and 7, it is shown that most of the energy of the LFM return echo is compressed into a narrower spike, which determines the target detection. The most important factor of radar target detection is to create the environment in which radar target is present. This includes simulation of the waveform along with the transmitter and receiver. For transmitting

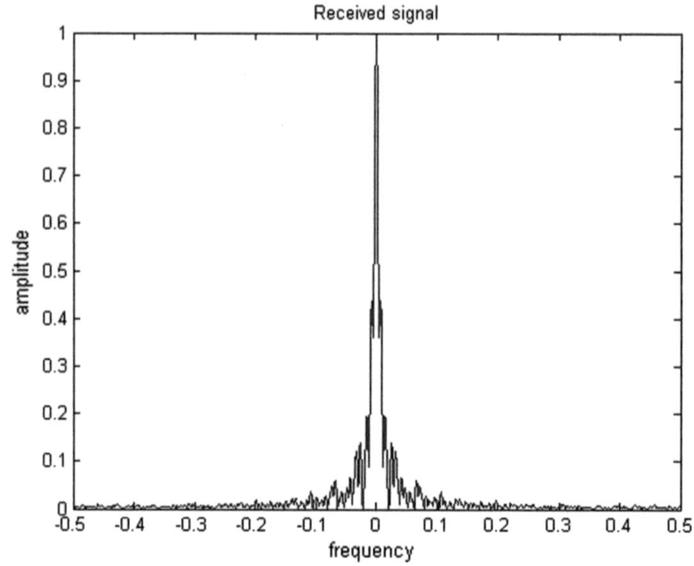

Fig. 6 Spectrum of reflected signal from Target1

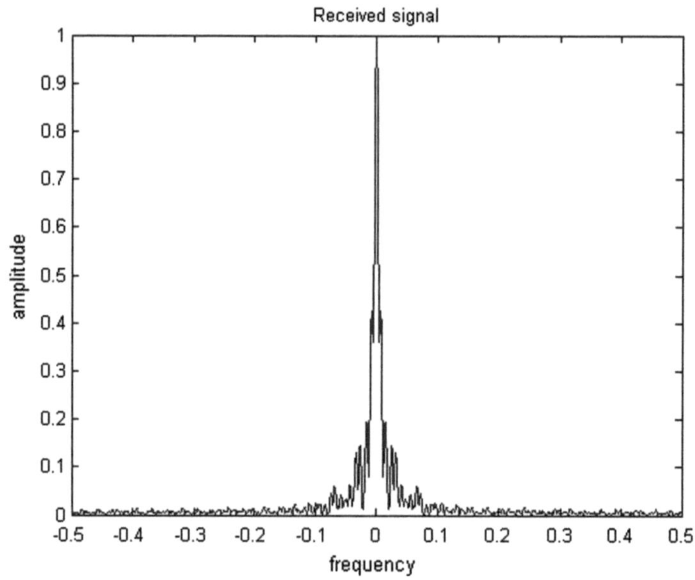

Fig. 7 Spectrum of reflected signal from Target2

these signals, a propagation environment is created considering different parameters. The radar signal is then transmitted to the direction of the simulated targets using a radiator, and then, the reflected echoes are collected using the collector. Analysis of the reflected signal is done from which some basic features are extracted.

5 Conclusion

ATR helps in finding an object in the presence of clutter, noise, etc. To extract information from the received signal so as to detect the target, the transmitting signal analysis should be done. In this case, we used a signal in THz band and its different characteristics are analyzed and used it as a transmitting signal. As low side-lobe level is essential for target detection, different side-lobe suppression techniques are implemented using different windows. If time and spectral analysis will not be there, then it will be difficult to extract the features related to the signal. The change in the frequency band when the signal coincides with the target will give us the idea about the target. By comparing the frequency band of the transmitted signal and received signal, we can make the decision about the target. The energy spectral density (spectral analysis of the signal) is the another important factor. Difference in which gives us the information related to the desired target. Therefore, time- and frequency-domain analysis of the signal is considered to be important for the RADAR target detection.

Acknowledgements The authors would like to acknowledge Assam Don Bosco University, Guwahati, for providing the resources necessary for compiling this paper.

References

1. Akyildiz IF, Jornet JM, Hana C (2014) Terahertz band: next frontier for wireless communications. Phys Commun 12:16–32
2. Kleine-Ostmann T, Nagatsuma T (2011) A review on terahertz communications research. J Infrared Millim Terahertz Waves 32(12):143–171
3. Tonouchi M New frontier of terahertz technology, Terahertz Photonics Laboratory, Institute of Laser Engineering, Osaka University
4. Kemp MC, Glauser A, Baker C (2006) Recent developments in people screening using terahertz technology-seeing the world through terahertz eyes. In: Proceedings of SPIE, vol 6212, pp 27–34
5. Tonouchi M (2007) Cutting-edge terahertz technology. Nat Photonics 1:97–105
6. Siegel PH (2002) Terahertz technology. IEEE Trans Microw Theory Tech 50:910–928
7. Shen X, Dietlein CR, Grossman E, Popovic Z, Meyer FG (2008) Detection and segmentation of concealed objects in terahertz images. IEEE Trans Image Process 17(12):2465–2476
8. Dudgeon DE, Lacoss RT (1993) An overview of automatic target recognition. The Lincoln Lab J 6(1)
9. Qiang F, Wenxian Y Automatic target recognition based on incoherent radar returns. IEEE 123–128

10. Andrews JR (2000) Picosecond pulse generators for UWB radars. Picosecond pulse labs, Boulder, CO, Application Note AN-9
11. Han J, Ngungen C (2004) Ultrawideband electronically tunable pulse generators. IEEE Microw Wirel Compon Lett 14(30):112–114
12. Rial FI, Lorenzo H, Pereira M, Armesto J (2009) Waveform analysis of UWB GPR antennas. Sensors 9:1454–1470. https://doi.org/10.3390/s90301454
13. Bhattacharyya K, Sarma KK (2002) Efficient signal processing for design of ANN based automatic target recognition (ATR) system. In: Proceeding of 57th annual technical session and journal of Assam science society, vol 50, Assam Science society
14. Shaika MU, Rao V (2012) Pulse compression techniques of phase coded waveforms in radar. Int J Sci Eng Res 3(8)
15. Skolnik MI (2002) Opportunities in radar. Electron Commun Eng J (ECEJ) 263–272. IEEE
16. Staderini EM (2002) Everything you always wanted to know about UWB radar: a practical introduction to the ultra wideband technology. http://67.225.133.110/gbpprorg/mil/radar/osee.pdf. Accessed 05 June 2015
17. Nathanson FA (1969) Radar design principles: signal processing and the environment. McGraw Hill, New Delhi
18. Gini F, Farina A, Greco M (2001) Selected list of references on radar signal processing, IEEE Trans Aerosp Electron Syst AES-37, 329–359

Detection of Endangered Gangetic Dolphins from Underwater Videos Using Improved Hybrid Frame Detection Technique in Combination with LBP-SVM Classifier

Swetha Sridharan, R. Dhaya, Kanthavel, Supraja and Sowjanya

Abstract Underwater surveillance is an emerging technology that helps overcoming the threats of aquatic world. These papers aim at identifying the endangered Gangetic dolphins using object recognition techniques and prevent them from extinction. We propose an improvised background subtraction and frame differencing algorithm to detect the presence of Gangetic dolphins. The improvisation includes iterating the hybrid algorithm N times to detect extremely slow-moving objects under poor illumination conditions. Also, the video is analysed and processed to classify if the object identified belongs to the endangered Gangetic dolphin class using the traditional SVM classifier combined with the powerful local binary pattern feature extractor. The algorithm solves the problems of object detection and recognition under poor illumination conditions. It showed improvements in terms of accuracy and performance in comparison with its previous versions. The proposed method can be applied aptly in the field of marine and aquatic object detection. They can be mainly applied to find objects exhibiting wavelike transitions, say fishes and other aquatic organisms with extreme accuracy and precision.

Keywords Frame differencing · Object detection and recognition

---

S. Sridharan (✉) · R. Dhaya · Supraja · Sowjanya
Department of Computer Science and Engineering, Rajalakshmi Engineering College, Chennai, India
e-mail: yesyes2311@gmail.com

R. Dhaya
e-mail: dhayavel2005@gmail.com

Supraja
e-mail: suprajabaskaran3@gmail.com

Sowjanya
e-mail: sowjanya.m.2015.cse@rajalakshmi.edu.in

Kanthavel
Department of Electronics and Communication Engineering,
Rajalakshmi Institute of Technology, Chennai, India
e-mail: kanthavel2005@gmail.com

S. Bhattacharyya et al. (eds.), *Recent Trends in Signal and Image Processing*,
Advances in Intelligent Systems and Computing 727,
https://doi.org/10.1007/978-981-10-8863-6_12

1 Introduction

Video image processing and underwater surveillance are an emerging field of research due to the presence of lots of unexplored secrets and facts of the underwater world. Their unique and marvellous habitat attracts lots of researchers to conduct study on their depleting wildlife and to protect and maintain its inhabitants in their natural environment. The vulnerability of their extinction is mainly due to human intervention and deposits of wastes in the water bodies. According to the International Union for Conservation of Nature (IUCN), the Platanista gangetica, also known as the Gangetic dolphin, is a critically endangered species [1]. They live only in freshwater ecosystems especially in the Gangetic belts of India. The real-time images of the Gangetic dolphins are shown in Fig. 1a–c [2]. They are found to be blind and hunt their prey by emitting ultrasonic sounds. The females are potentially larger than the males. The female dolphins give birth to only a single calf once in three years. Hence, the species is critically endangered. Moreover, they are in verge of extinction due to various factors like water development projects, mortality due to fishing gears like gillnets [3, 4]. There is an urge in detecting and protecting them from protection. The extinction of these endangered species of mammals may in turn cause an imbalance in the ecosystem. The extinction of these species may also have severe negative effects on the other aquatic organisms. The manual approach for protecting them is time-consuming and may lead to great disasters in case of delay in detection. Hence, automating the process of protecting the wildlife is of a prime importance. This paper proposes an algorithm to detect the endangered species automatically using the approaches of background subtraction and frame differencing. The existing background subtraction and frame differencing techniques suffer from the problems of depleted performance in poor illumination. Therefore, the modified algorithm proposes a hybrid approach to overcome the illumination issues. Also, the hybridization includes features to detect objects showing wave transitions very effectively. Dolphins and fishes tend to show locomotion by swimming, which can be defined as a wavelike motion using the tail and its fins [5, 6]. This improved hybrid frame differencing algorithm proposed in this paper also provides a special focus on detection of objects showing wavelike motion. It also removes the challenges faced by existing system like detecting extremely slow-moving objects. The proposed methodology uses the underwater surveillance video data set as the source of data. As a next step, the videos are converted into image frames and processed to remove noise from the converted image data. The proposed algorithm for detecting objects showing wavelike motion is applied to the data to detect the endangered Gangetic dolphins from the wide variety of the organisms in the images. A system comprising of the SVM classifier is trained by providing the images of endangered species of Gangetic dolphins under consideration. When the output of the improved hybrid frame differencing detection is provided to the classifier, the SVM returns if the detected underwater object at the location belongs to the endangered Gangetic dolphin class or not. This can be used by the coastal and marine guards to save them and take necessary steps to persist their existence.

Fig. 1 Real-time images of critically endangered Gangetic dolphins

2 Literature Survey

Sfakiotakis et al. [6] discussed a few physic mechanical plans which advanced in fish moving purposes in sub merged vehicles. The fish swimming sorts were displayed, taking after the entrenched arrangement plan and classification initially proposed by Breder Angle, swim either by body as well as caudal blade (BCF) developments or utilizing middle or potentially matched balance (MPF) drive. The last was for the most part utilized at moderate rates, offering more prominent mobility and better propulsive proficiency, while BCF developments can accomplish more noteworthy push and increasing speeds. For both BCF and MPF motions, particular swimming modes are distinguished, in the light of the propeller and the sort of developments utilized for push era. Alongside general portrayals and kinematic information, the explanatory methodologies created to concentrate each swimming mode are additionally presented. Specific reference is made to lunate tail drive, undulating balances and labra form (oscillatory pectoral blade) swimming components, distinguished as having the best potential for abuse in manufactured frameworks. Chavan et al. [7] talked about a coordinated way to deal with recognize shot limit utilizing histogram contrast of progressive casings in a video arrangement with the accentuation of twin correlation strategy and monochrome casing discovery is exhibited. The system could separate unexpected and progressive moves and gave a powerful calculation to work in an execution of shot limit recognition. He has provided a clear picture for short and slow moving objects. Viriyakijjaa and Chinnarasria [8] concentrated on the improvement of picture preparing strategies for wave estimation in a wave flume as an apparatus to supplant conventional wave gages, which were muddled to utilize, expensive and intrusive. The exploratory outcomes demonstrated the understanding of water surface rise from both estimation techniques. The normal wave tallness and normal wave time of consistent wave acquired from the zero up-intersection investigation were distinctive by under 10 and 1% individually contrasted and customary wave gages. Mahmoud Abdulwahab et al. [9] examined a portion of the systems which is fluctuating from straightforward strategies, for example outline differencing and guess middle channel, to more muddled probabilistic demonstrating procedures. The outcomes demonstrated that straightforward methods, for example estimate middle channel, can create great outcomes with much lower computational multifaceted nature. Alex and Wahi [10] proposed another BSFD calculation which redressed the detriment of the foundation subtraction technique and the edge distinction strategy and proposed a dynamic refreshing of foundation picture by casing contrast strategy and used the force of the foundation subtraction strategy for recognizing the moving article successfully and precisely. Chavan et al. [11] proposed an algorithm to dissolve detection. In this calculation, shading histogram contrast between successive edges was figured and normal estimation of this distinction for every single continuous casing was utilized as a metric for disintegrating identification. Singla [12] displayed a new calculation for identifying moving items from a static foundation scene in the light of edge distinction. The main casing was caught through the static camera, and after that, grouping of edges was caught at customary interims. Furthermore, the total distinction was figured between the back-to-back edges and the distinction picture put away in the framework. Thirdly, the distinction picture was changed over into dim picture and after that converted into paired picture. At last, morphological separating was done to evacuate clamour. Mahapatra et al. [13] proposed a multi-threading fluffy c-mean (MFC) way to deal with recognize multimoving fishes in a boisterous and thick condition. In that approach, the writers joined the multistrung parallel (MTP) approach and part-based approach for optical stream. A fluffy c-mean idea gave as a supporting element. The re-enactment comes about demonstrated that the proposed technique can ready to track and distinguish submerged fishes with high recognition rate. Liu et al. [14] proposed an approach combining background subtraction and three-frame difference. In that method, firstly moving object pixels by background subtraction and three-frame difference were detected, respectively. Lastly, morphology processing was utilized on the result to remove the noise caused by non-static objects in background.

3 Proposed Method

The proposed system focuses on detecting the endangered species from the video data for underwater surveillance using the hybrid algorithm for detecting objects showing wavelike motion. The improved hybrid frame differencing algorithm is used for the object recognition process. The block diagram of the proposed methodology is shown in Fig. 2. The proposed method comprises of the following modules.

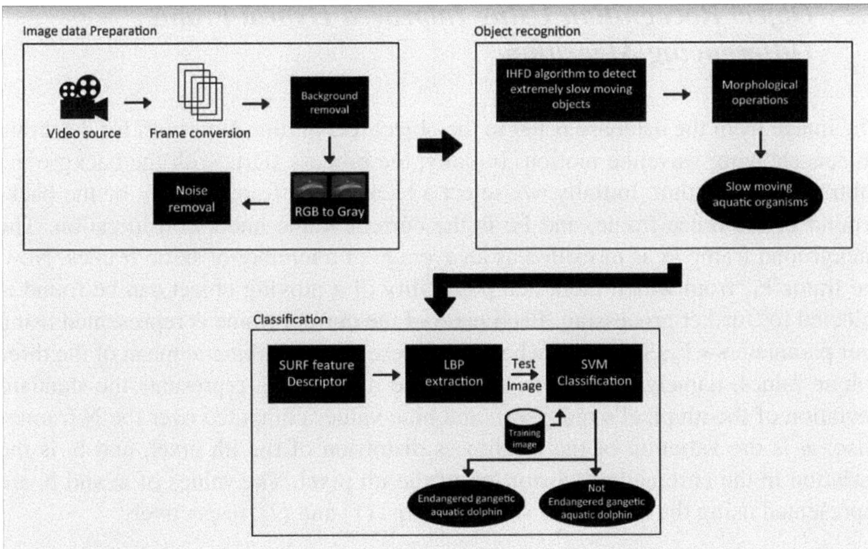

Fig. 2 Proposed system

3.1 Image Data Preparation

In this module, the underwater videos are recorded and initially converted into consecutive frames for further processing. Underwater explorations lack illumination and therefore require artificial illumination. This may in turn lead to changes in the brightness patterns as we move up the seabed. Hence, these patterns are to be removed [10, 15]. Further, the videos are converted to frames f_k where k is the number of the frames at a rate of 5 frames per second. The underwater video may suffer from problems of presence of artefacts like lens glare and electronic noise from the camera. Hence, the background is removed to ensure that the objects of interest are retained and help in better analysis. Also, the slow-moving objects alone are retained, and the fast-moving objects are ignored from the scene by using the background removal process. In order to ensure removal of Gaussian noise, the Gaussian filter is applied to the images obtained as frames. Now, the image data from the previous step are noise-free and ready for further processing.

3.2 *Object Recognition Using Improved Hybrid Frame Differencing Algorithm*

The image from the database is fed to the object recognition algorithm for detecting objects showing wavelike motion. Initially, the process starts with the background subtraction algorithm. Initially, we select a background frame. Let F_b be the background or reference frame, and Fc is the current frame under consideration. The background frame F_b is modelled as an average of a number of static frames. Now, the frame F_r, from which there is a possibility of a moving object can be found is selected for further processing. Each pixel of the current frame is represented using four parameters $< E_i, S_i, a_i, b_i >$ where E_i represents the arithmetic mean of the three colour values, namely red, green, blue of the ith pixel. S_i represents the standard deviation of the ith pixel's red, green and blue values computed over the N frames. Also, a_i is the variation of the brightness distortion of the ith pixel, and b_i is the variation in the chromaticity distortion of the ith pixel. The values of a_i and b_i are represented using the formula as shown in Eqs. (1) and (2), respectively.

$$a_i = Root\,Mean\,Square\,(a_i) = \sqrt{\frac{\sum_{i=0}^{N}(a_i - 1)^2}{N}} \tag{1}$$

$$b_i = Root\,Mean\,Square\,(CD_i) = \sqrt{\frac{\sum_{i=0}^{N}(CD_i)^2}{N}} \tag{2}$$

In Eqs. (1) and (2), the variable N is the number of frames considered. In Eq. (1), α represents the pixel's strength of brightness with respect to the expected value. The value of a_i is less than 1 if it is darker, 1 if the brightness of the pixel in the current image is the same as in the reference image and greater than 1 if it brighter. In Eq. (2), CD is the distance between the observed colour and the expected chromaticity line. The current frame obtained is subtracted from the reference frame. We choose a threshold value T to determine the rate of detection. The values of T can be automatically selected using the histogram of the background image F_b. Based on the threshold values, it is possible to detect if the detected pixel belongs to the moving object or the background. This can be represented using the formula given by Eq. (3)

$$|F_b - F_r| > T \tag{3}$$

In Eq. (3), F_r is the current frame under consideration. If the difference in pixel is greater than T, then the detected object is considered to be a foreground object. By using the background subtraction method, the process of detecting the moving objects is accomplished. This background subtraction method helps in overcoming

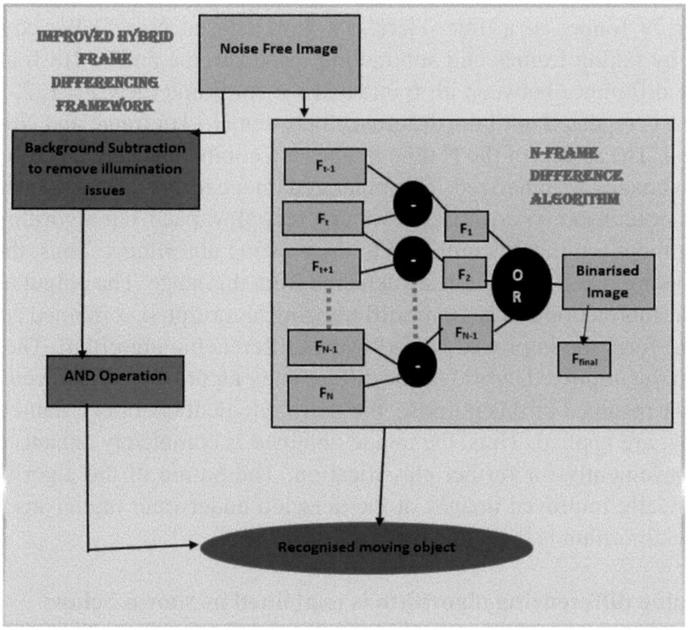

Fig. 3 Improved hybrid frame differencing framework

the illumination issues. Simultaneously, the N-frame differencing techniques are applied to improve the accuracy of the output. The frame differencing algorithm has the following advantages: (i) using the frame differencing method, the system provides solution to automatically detect objects even when the background is not static. (ii) The algorithm adapts rapidly to changing illumination conditions. (iii) The algorithm helps to overcome the problems of ghostly images. The improved frame differencing algorithm is best suitable for detecting moving fishes in the underwater videos as they show a steady and slow motion. Let F_t be the current frame, F_{t-1} be the previous frame, and F_{t+1} is the next frame under consideration where t = 1, 2, 3, …. The frame F_t from the output of the background subtraction is chosen. The difference is computed between two consecutive frames, and the result is stored in F_{result}. The output is combined to form the final differencing image containing the moving object. This can be represented using the formula (4).

$$F_{result} = F_t - F_{t-1} \tag{4}$$

$$F_{final} = F_{result(i)}\ (OR\ Operation)\ F_{result(i)} \forall_{i=1} \tag{5}$$

The final values of the frame differencing can be represented using F_{final} and can be represented by Eq. (5). In Eq. (5), and the value of N represents the number of iterations, which can be selected on a trial-and-error basis. The improvisation from the previously existing algorithms includes iterating the frame differencing method considering N frames at a time. Here, we have chosen N = 5. We consider the difference by taking frames and subtracting the ith frame and i−1th frame. Next, we find the difference between ith frame and i + 1th frame (here i = 1, 2, 3, …, N). The process is repeated until the difference between N−1th frame and Nth frame is encountered. The results of the N time images are combined using an OR operation. Finally, the images are binarized. This binarized image forms the resultant image. As the aquatic or underwater creatures move at a very slow pace, our algorithm helps in identifying them accurately compared to the existing algorithms. Thus, the moving objects showing wavelike motion are detected from this stage. The output of both the background subtraction and N-frame differencing algorithm is combined using AND operation to form the improved hybrid frame differencing algorithm. The resultant image from the improved hybrid frame differencing algorithm is enhanced further to obtain better results. For this purpose, the morphological operators, namely erosion and dilation, are applied. Thus, the image obtained is completely enhanced and can be used conveniently for further classification. The output of the algorithm is the morphologically improved images of the detected underwater organisms. The flow chart of the algorithm is depicted in Fig. 3.

The N-Frame differencing algorithm is explained as shown below:

Algorithm: *N- Frame differencing algorithm*
Assumptions: *Let I be the frame image under consideration and I -1 be the previous frame image. Also,* I_f
is the resultant image. Let w be the size of the window in the image. Let N=5.
Input: *Noise free images from underwater video sliced at the rate of 1ms.*
Output: *I final is the final output image containing the recognised moving object.*
Start
For each image obtained at an interval of 1ms from the underwater video,
Choose a value of N.
Do

```
        For each pixel p_i in the current window (w) of image I,
          Do
             Pixel value = p_i (I-1)- p_i (I);
             If (Pixel value ==0)
                   {Set p_i (I_f) = 0;}
             Else
                   {Set p_i (I_f) =1;}
             I=i+1;
          Until (The window covers the entire image)
            f=f+1
            I_final= I_f (OR Operation) I_f+1;
Until (f=N-1)
Until (f=N-1)
```

/ By doing so the pixels between the frames which differ are highlighted. If they are highlighted, it implies that there is a movement in the object. The* I_{final} *is the resultant of OR operation between the two frames. The do loop is proceeded until N times to show extreme slow movements. */*
End.

4 Classification of Endangered Gangetic Dolphins Using Local Binary Patterns and Support Vector Machines

The next stage is to classify the image as the specified endangered species, i.e. Gangetic dolphin. For this, we use the image classification techniques. The following are the steps to be followed for the classification process. As the moving object-detected frame is noise-free, it does not require any pre-processing steps. The first step involves extracting features from the image. For this purpose, we use the combination of speeded up robust features [SURF] and local binary pattern [LBP-SURF descriptor with colour invariant and texture-based features for underwater images]. In this method, the points of interest are detected using Hessian-based feature detection method. LBP starts by dividing the underwater image into windows of appropriate dimensions (say 5×5). Consider the centre most pixel of the current window and follow a circular path to compare the centre pixel with each of its neighbour pixels, say p_i. If the value of the centre pixel is greater than its neighbour value, we replace the value of the centre pixel as 1 otherwise 0. The pixel p_i is compared with its neighbour p_{i+1} along the traced circular path. Again, the value of the pixel p_i is replaced with a 1 if it is greater than p_{i+1} or 0 if it has a lesser value. By proceeding in this fashion, the entire image is binarized and we get the LBP image [16]. The final stage is matching the descriptors to decide if this point belongs to the object of interest or not. The matched points can be used for further processing and performing underwater object recognition. Now, we use the traditional SVM classifier to classify if the underwater object is a Gangetic dolphin or not. Support vector machine (SVM) classifier can be used to classify both linear and nonlinear data. It deals with the problem of classifying data by transforming the original training data into a higher dimension using a nonlinear mapping. SVM constructs a hyperplane which separates the image such that each image at the maximum possible distance from the closest resembles image in the training set. SVM finds the optimal hyperplane from the set of hyperplanes. An optimal hyperplane or maximal margin hyperplane is one which separates the data effectively such that there is maximum separation between the classes. SVM finds the hyperplane using the support vectors and margins [17].

$$\mathrm{w} \cdot \mathrm{x} + \mathrm{b} = 0 \tag{7}$$

Equation (7) represents the equation of the hyperplane. In Eq. (7), w is the weight vector, x is the training vector, and b is the bias. The bias b can be replaced by an additional weight term w_0, and the Eq. (7) can be rewritten as shown in Eq. (8) as follows:

$$w_0 + w_1x_1 + w_2x_2 = 0 \tag{8}$$

$$w_0 + w_1x_1 + w_2x_2 > 0 \tag{9}$$

$$w_0 + w_1x_1 + w_2x_2 < 0 \tag{10}$$

Any value above the separating hyperplane satisfies Eq. (9), and any value below the separating hyperplane satisfies Eq. (10). SVM operates in two phases, namely the training and testing phases. The training phase takes all the input images from the data set and learns about the images based on the features present in them. In the next phase, the testing images are supplied and the images are classified into the appropriate classes. SVM-based classification involves two stages. In the first or training phase, the classifier is trained using the training image data set containing the images of around 50 images. Now, the second or testing stage involves providing the image from which the underwater object is to be classified as a Gangetic dolphin or not. The classifier compares the features and returns the output. The output of the module is revealing if the image belongs to the class of endangered Gangetic dolphins or not.

5 Experiment

The experiment was conducted using the proposed system called improvised background subtraction and frame differencing algorithm called (IBSFD). The experiment was aimed to detect critically endangered Gangetic dolphins from underwater sequences. These video sequences are sampled to 320×240 with a 24-bit RGB and at a frame rate of 5 frames per second. Each video of the data set is about 1 minute long (300 frames). The aim of this experiment is to determine the accuracy of the detection of the proposed slow-moving underwater object recognition algorithm and counting abilities that make use of the underwater object tracking modules (Fig. 4).

The proposed algorithm for detecting objects showing wavelike motion is applied to the data to detect the dolphins from the wide variety of the organisms in the images. A system comprising of the SVM classifier is trained by providing the images of endangered species of mammals (Gangetic dolphins) under consideration. When the output of the improved hybrid frame differencing detection is provided to the classifier, the SVM returns if the object detected at the location belongs to the endangered Gangetic dolphin class or not. This can be used by the coastal and marine guards to save them and take necessary steps to persist their existence.

Fig. 4 Segmented images of critically endangered Gangetic dolphins

6 Conclusion

In this paper, we have proposed an improvised background subtraction and frame differencing algorithm called (IBSFD) to recognize the nearness of underwater moving objects from the submerged recordings. The proposed algorithm empowered us to distinguish the articles under poor light conditions and gave the successful recognition of question moving in wave shape too. The outcomes are broken down and prepared to characterize if the mammal recognized is nearly termination or not. With the assistance of the customary SVM classifier, the algorithm has tackled the issues of protest discovery and acknowledgment under poor light conditions. The predetermined outcomes demonstrated that the algorithm performed better for discovery of items displaying wavelike moves, say angles and other oceanic creatures.

References

1. IUCN red list (2017). http://www.iucnredlist.org/details/41758/0
2. Ganges river dolphin. https://www.worldwildlife.org/species/ganges-river-dolphin
3. Dolphin—features. https://en.wikipedia.org/wiki/South_Asian_river_dolphin
4. Indian red data list. https://www.pmfias.com/iucn-red-list-india-red-data-list-red-book/
5. Fish locomotion (2017). https://en.wikipedia.org/wiki/Fish_locomotion
6. Sfakiotakis M, Lane DM, Davies JB (1999) Review of fish swimming modes for aquatic locomotion. IEEE J Oceanic Eng 24(2):237–252
7. Chavan S, Akojwar S (2016) An efficient method for fade and dissolve detection in presence of camera motion and illumination. In: Proceedings of international conference electrical, electronics, and optimization techniques (ICEEOT), pp 3002–3007
8. Viriyakijja K, Chinnarasri C (2015) Wave flume measurement using image analysis. In: Proceedings of aquatic procedia, pp 522–531
9. Mahmoud Abdulwahab A, Khalifa OO, Rafiqul Islam MD (2013) Performance comparison of background estimation algorithms for detecting moving vehicle. World Appl Sci J 21, 109–114
10. Alex DS, Wahi A (2014) BSFD: background subtraction frame difference algorithm for moving object detection and extraction. J Theor Appl Inf Technol 60(3)
11. Chavan SA, Alkari AA, Akojwar SG (2014) An efficient method for gradual transition detection in presence of camera motion
12. Singla N (2014) Motion detection based on frame difference method. Int J Inf Comput Technol 4(15):1559–1565
13. Mahapatra SK, Mohapatra SK, Mahapatra S, Tripathy SK (2016) A proposed multithreading fuzzy c-mean algorithm for detecting underwater fishes. In: Proceedings of 2nd international conference on computational intelligence and networks (CINE), pp 102–105
14. Liu H, Dai J, Wang R, Zheng H, Zheng B (2016) Combining background subtraction and three-frame difference to detect moving object from underwater video. In: Proceedings of OCEANS 2016, Shanghai,10 Apr 2016, pp 1–5
15. Rzhanov Y, Huff LC, Cutter Jr RG (2002) Underwater video survey: planning and data processing
16. Sridharan Swetha, Angl Arul Jothi V, Rajam Mary Anita (2016) Segmentation of tumors from brain magnetic resonance images using gain ratio based fuzzy c-means algorithm. Int J Comput Sci Inf Technol 7(3):1105–1110
17. Han J, Kamber M, Pei J (2012) Data mining: concepts and techniques. Elsevier

Automatic Lip Extraction Using DHT and Active Contour

Amiya Halder and Souvik Dutta

Abstract Lip contour detection and extraction is the most important criterion for computerized speech reading. In this paper, lip contour extraction algorithm is proposed based on active contour model across the discrete Hartley transform (DHT). This approach has efficiently solved the problem of real-time lip detection. Most of the existing lip extraction methods impose several restrictions, but this algorithm tries to approach an unconstrained system. This method works good in case of uneven illumination, effects of teeth and tongue, rotation, deformation. This proposed algorithm experiments on large number of lip images and gives the satisfactory performance than other existing lip extraction methods.

Keywords Image segmentation · Discrete Hartley transform · Active contour
Illumination compression

1 Introduction

In videoconferencing, lip-reading or low-bit rate communication, speech recognition and other multimedia systems, lip segmentation is essential. Hence, it is required to develop a more accurate and robust method for extracting lip contours automatically for lip-reading purpose [1–6]. The presence of great variations in images caused by speakers' utterances and lighting conditions is the constraint of lip region detection, and also the low chromatic and luminance contrast between the lip feature and the remaining face skins for unadorned face is a problem for lip segmentation.

The lip detection approaches are mainly two ways: edge-based approach as well as model-based approach. Lip segmentation methods use segmentation of lip directly from colour space by applying colour filter or colour transformation on the images to

A. Halder (✉) · S. Dutta
STCET, Kidderpore, Kolkata 700023, West Bengal, India
e-mail: amiya.halder77@gmail.com

S. Dutta
e-mail: duttasouvik654@gmail.com

S. Bhattacharyya et al. (eds.), *Recent Trends in Signal and Image Processing*,
Advances in Intelligent Systems and Computing 727,
https://doi.org/10.1007/978-981-10-8863-6_13

make the colour differences between the lip and other part of the face [7]. The time required to process these algorithms is less. But low contrast between the lip hue and the remaining face creates a major problem [8]. Since the lip extraction is very much dependable on colour change, these methods cannot satisfactorily extract the boundary of lip. This problem is solved by the clustering of lip regions that allows segmentation using differences in the colour between the mouth regions and the remaining face regions [9, 10]. The number of clusters is provided in this method. In some algorithms, the lip region is estimated depending on colour intensity ratio [11]. After that, some morphological operation is done on the resulting image to make the lip portion accurate and prominent. Then for lip region extraction, curve fitting technique is applied, but for weak contrast of face region, lip region cannot be thresholded properly.

Wavelet transform is another important technique for edge detection. This can be applied directly for the detection of lip contours, but this method does not give good result because of weak colour difference between the labial contour or outline of lip and its skin background. Guan has approached wavelet edge detection across the chrominance component of DHT [12].

Another important lip extraction approach is model based approach like deformable template (DT), active shape model (ASM), active contour model (ACM) [13], localized active contour model (LACM) [14]. Among these techniques active contour model or snake has a lot of application in computer vision, object tracking, shape recognition.

In this paper, a novel automatic lip contour extraction algorithm is proposed based on the geodesic active contour model (GACM) [15] across the chrominance component of DHT. Geodesic approach for segmentation of object relates the conventional snake model based on minimization of energy and geometric active contour based on theory of curve evaluation. Here, the change in the chrominance component to select the differently coloured rather than naturally toned lip portion is described in detail. The performance of this proposed method is better than existing algorithms.

2 Geodesic Active Contour on DHT

In this paper, chrominance component of DHT is used for the betterment of the contrast effect between the lip and the skin area. Lip region is extracted by applying geodesic active contour model on chrominance component.

2.1 DHT

DHT is an invertible linear transform of discrete, periodic data like discrete Fourier transform. But in DFT, each input is multiplied by $(\cos(\omega) - i * \sin(\omega))$, but in DHT, each input is multiplied simply by $(\cos(\omega) + \sin(\omega))$. In this paper, the DHT is carried out as follows [12] for normal lip tone:

$$X = \frac{1}{\sqrt{\rho}} \begin{pmatrix} 1 & 1 & 1 \\ 1 & cas\left(\frac{2\pi}{\rho}\right) & cas\left(\frac{2\pi(\rho-1)}{\rho}\right) \\ 1 & cas\left(\frac{2\pi(\rho-1)}{\rho}\right) & cas\left(\frac{2\pi(\rho-1)^2}{\rho}\right) \end{pmatrix} \tag{1}$$

For violet-coloured lip i.e. if violet-coloured lipstick is added to lip, the DHT is like as follows:

$$\widehat{X} = \frac{1}{\sqrt{\rho}} \begin{pmatrix} 1 & 1 & 1 \\ 1 & cas\left(\frac{2\pi}{\rho}\right) & cas\left(\frac{2\pi(\rho-1)}{\rho}\right) \\ cas\left(\frac{2\pi(\rho-1)^2}{\rho}\right) & cas\left(\frac{2\pi(\rho-1)}{\rho}\right) & 1 \end{pmatrix} \tag{2}$$

where $cas(\omega) = \cos(\omega) + \sin(\omega)$, and X and $\widehat{X}$ are the convolution matrices. For $\rho = 3$ (RGB components), Eqs. 1 and 2 can be written as

$$X = \begin{pmatrix} 0.5773 & 0.5773 & 0.5773 \\ 0.5773 & 0.2113 & -0.7886 \\ 0.5773 & -0.7886 & 0.2113 \end{pmatrix} \tag{3}$$

$$\widehat{X} = \begin{pmatrix} 0.5773 & 0.5773 & 0.5773 \\ 0.5773 & 0.2113 & -0.7886 \\ 0.2113 & -0.7886 & 0.5773 \end{pmatrix} \tag{4}$$

After convoluting Eq. 3 with R, G, B components

$$\begin{pmatrix} X_1 \\ X_2 \\ X_3 \end{pmatrix} = \begin{pmatrix} 0.5773 & 0.5773 & 0.5773 \\ 0.5773 & 0.2113 & -0.7886 \\ 0.5773 & -0.7886 & 0.2113 \end{pmatrix} \begin{pmatrix} R \\ G \\ B \end{pmatrix} \tag{5}$$

and after convoluting Eq. 4 with each of R, G, B components, we get

$$\begin{pmatrix} \widehat{X_1} \\ \widehat{X_2} \\ \widehat{X_3} \end{pmatrix} = \begin{pmatrix} 0.5773 & 0.5773 & 0.5773 \\ 0.5773 & 0.2113 & -0.7886 \\ 0.2113 & -0.7886 & 0.5773 \end{pmatrix} \begin{pmatrix} R \\ G \\ B \end{pmatrix} \tag{6}$$

We consider $Z_i = X_i$ or $\widehat{X_i}$ where $i \in [1, 3]$. $Z_1 Z_2 Z_3$ has two major components: luminance (Z_1) and chrominance ($Z_2 Z_3$). Luminance (Z_1) is characterized as the brightness of the colour pixels, and chrominance ($Z_2 Z_3$) illustrates the colour saturation. But It has been observed that the chrominance component (Z_3) of ($Z_2 Z_3$) consists of high value around the lip.

2.2 Geodesic Active Contour

In this paper, geodesic active contour is used for lip detection on the X_3 component [14]. This approach depends on active contour evolving in time according to intrinsic geometric measures of the image.

Classical snake model: Let $G(q) : [0, 1] \rightarrow \Re^2$ be a parameterized planar curve, and let $I : [0, a] \times [0, b] \rightarrow \Re^+$ be a given image in which a particular object is to be detected. The classical snake approach is related to the curve G with an energy. The equation of this is given by

$$E(G) = \alpha \int_0^1 |G^{/}(q)|^2 dq + \beta \int_0^1 |G^{//}(q)|^2 dq - \gamma \int_0^1 |\nabla I(G(q))| dq \tag{7}$$

In which α, β and γ are the real positive constants. The curve smoothing can be obtained by considering $\beta = 0$ in Eq. 7 only having the first regularization term, which is given by

$$E(G) = \alpha \int_0^1 |G^{/}(q)|^2 dq - \gamma \int_0^1 |\nabla I(G(q))| dq \tag{8}$$

For minimizing Eq. 8, $-|\nabla I|$ in (8) can be replaced by $f(|\nabla I|)^2$; thereby, the general equation is given by

$$E(G) = \alpha \int_0^1 |G^{/}(q)|^2 dq + \gamma \int_0^1 f(|\nabla I(G(q))|)^2 dq \tag{9}$$

$$= \int_0^1 (E_{int}(G(q)) + E_{ext}(G(q))) dq \tag{10}$$

Geodesic active contours can be derived from Eq. 8. But the functional in Eq. 8 is not intrinsic because it depends on arbitrary parameter q. With the help of parameterized representation of the functional in Eq. 8, the curve via $q = \phi(r), \phi : [a, b] \rightarrow [0, 1], \phi^{/} > 0$, from that it is obtained

$$\int_0^1 |G^{/}(q)| dq = \int_a^b |(G \circ \phi)^{/}(r)|^2 (\phi^{/}(r))^{-1} dr \tag{11}$$

$$\int_0^1 f(|\nabla I(G(q))|) dq = \int_a^b f(|\nabla I(G \circ \phi(r))|) \phi^{/}(r) dr \tag{12}$$

3 Proposed Algorithm

The steps of the proposed algorithm are as follows.

3.1 Illumination Compression

Different lighting conditions can vary the appearance of the skin colour to the human eyes. In different lighting conditions, to enhance the lip detection efficiently, illumination compression is required before applying Eqs. 5 and 6. For this purpose, we apply the log function of the intensity [12]. The log transformation is used to normalize the luminance or brightness level which is given by

$$\xi(\mu, \nu) = \delta + \frac{\log(\zeta(\mu, \nu) + 1)}{\eta \log(\theta)} \tag{13}$$

where $\zeta(\mu, \nu)$ is the original image and δ, η and θ are the constant values or parameters that control the shape and location of the curve. Perform this illumination compression on the original image $\zeta(\mu, \nu)$, and normalize the resulting image $\xi(\mu, \nu)$ in the range of [0, 255] (normalize each component of the resulting image). This equation is applicable for any value of δ, η and θ except $\beta = 0$ and $\theta = 1$ or $\theta \leq 0$. In this experiment, the values of δ, η and θ are chosen to be 10, 0.25 and 2, respectively.

3.2 Lip Segmentation

After illumination compression, the components of the processed image are transformed into $Z_1Z_2Z_3$ as mentioned in Eqs. 5 and 6. Z_3 component is directly normalized in the range of [0, 255] to increase the contrast between the lip area and the connected skin area using histogram stretching. Then, specify the mask around the lip region as a binary image having the same size as the original face image. Also, it needs to specify the maximum number of iterations to perform the evolution of the segmentation. Active contour stops when the result of the current iteration matches with previous five contour positions.

3.3 Lip Boundary Detection

The resulting image after applying active contour is the binary image representing segmented lip region. Then, draw boundaries of regions in the binary image on the current axes. In binary image, the foreground and background regions are represented by 1 and 0 respectively. So, by using binary image boundaries, lip boundary pixel locations in the original image are found. In this way, the outline of the lip contour is found in the original image.

4 Experimental Results

To test and verify the performance of the proposed method of lip segmentation, several experiments are performed on different sizes of colour images taken from IMM face database [16]. The results of this segmentation algorithm are compared with those of the following methods: multi-scale wavelet edge detection (MSWE) algorithm [12], spatial fuzzy clustering (SFCA) algorithm [9], syntactic approach (SA) algorithm [2], morphological reconstruction-based segmentation (MR) algorithm [11] shown in Fig. 1. To show the comparison result, different lip images with various shapes and oral opening are considered. From these segmentations, it is easily understood that SFCA cannot extract lower contour and lower portion of the lip exactly. Also, in this method, other parts of the images are also extracted along with

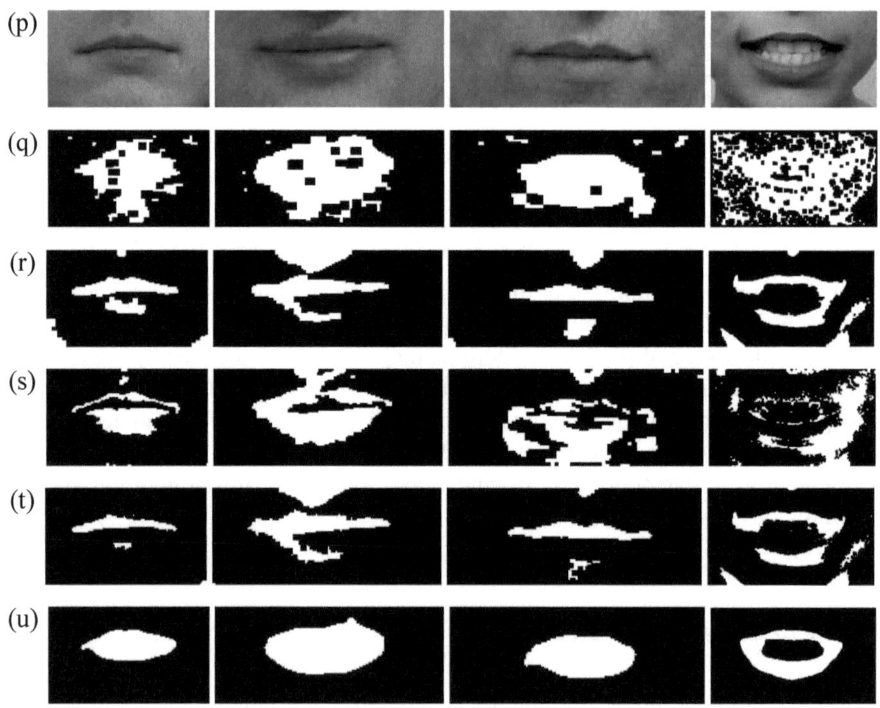

Fig. 1 Comparison of extracted lip regions using **q** MSWE, **r** MR, **s** SA, **t** SFCA, **u** proposed method and **p** the original lip images from IMM face databases

Table 1 Comparison of accuracy measure lip image using MSWE, MR, SA, SFCA and proposed method

	Lip 1	Lip 1	Lip 2	Lip 2	Lip 3	Lip 3	Lip 4	Lip 4	Lip 5	Lip 5	Lip 6	Lip 6
	F_{se}	F_{ol}	F_{se}	F_{ol}	F_{se}	F_{ol}	F_{se}	F_{ol}	F_{se}	F_{ol}	F_{se}	F_{ol}
MSWE	53.23	75.39	63.53	49.59	68.79	48.29	69.99	39.68	76.70	29.32	32.96	161.57
SFCA	60.04	43.88	53.37	42.70	52.44	41.67	46.92	46.59	58.54	43.88	35.47	87.21
SA	54.77	47.99	72.38	30.67	47.22	64.83	70.45	40.96	46.86	60.19	8.02	171.15
MR	56.19	33.41	57.94	38.57	55.06	34.50	52.82	41.83	61.17	49.00	31.18	134.88
PropMeth	87.72	13.16	91.08	9.57	90.00	10.01	87.21	14.23	96.90	3.08	90.03	10.02

the lip contour. In case of MR method, the same case arises. This is a serious problem as far as lip contour segmentation is concerned. From SA method, the idea of the shapes of the lip edges is got but it does not provide the lip contour which is exactly filled. In MSWE method, the regions where the color contrast is very low cannot be extracted properly. This proposed method can give the correct idea of the shapes of the lips, and it can segment the exact lip region. A comparative study with the five methods MSWE, SFCA, SA, MR and proposed method on different lip images is given in Table 1. From that, it can be easily understood that the proposed method works efficiently and provides superior performance even in the challenging situations. Also, in Fig. 2, we have shown average accuracy value of F_{ol} and F_{se} on 150 lip images. It is shown that our proposed method gives better results than other existing method in case of normal lip or violet-coloured lip images.

$$F_{ol} = 2 \times \frac{(P_1 \cup P_2)^*}{P_1 + P_2} \times 100 \tag{14}$$

where P_1 and P_2 are represented as segmented lip region and ground truth, respectively. The segmentation error (F_{se}) is measured by [10]

$$F_{se} = \frac{F_{ole} + F_{ile}^*}{2 \times TL} \times 100\% \tag{15}$$

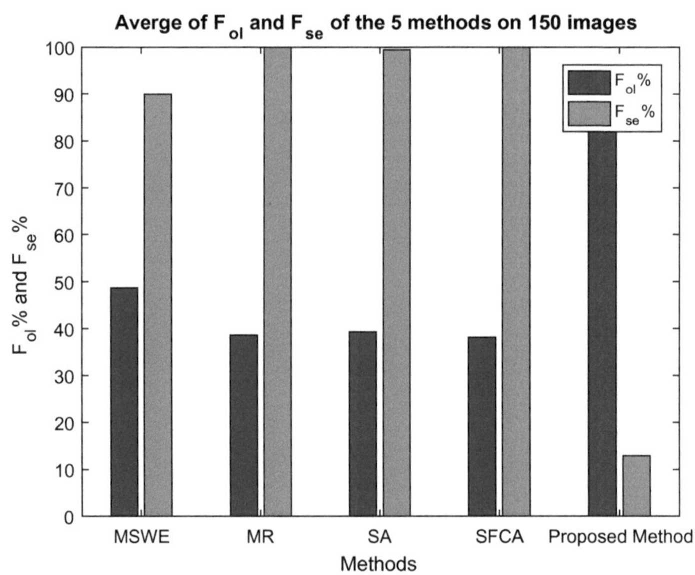

Fig. 2 Comparison of accuracy measure average of 150 lip images using MSWE, MR, SA, SFCA and proposed method

where F_{ole} is the number of non-lip pixels (outer lip error) and F_{ile} is the inner lip pixels. TL signifies the ground truth pixels. Total agreement will have an F_{se} of 0%.

5 Conclusion

Constructive automatic lip extraction from colour images is sometimes difficult because of weak colour contrast between lip and the remaining skin region. In this paper, lip detection and segmentation is proposed based on geodesic active contour model across the chrominance component of DHT. This algorithm can satisfactorily exploit the spatial interactions between neighbouring pixels by using geodesic active contour model across chrominance component. The proposed algorithm segments the lip region with greater accuracy without any requirement of optimum threshold for each lip image. In this algorithm, geodesic active contour model minimizes the segmentation error. From the comparative study with MSWE, SFCA, SA and MR, it is clear that the proposed algorithm results in a superior performance in segmenting lip region even if the lip is externally toned with different colours.

References

1. Sadeghi M, Kittler J, Messer K (2001) Real time segmentation of lip pixels for lip tracker initialization. In: Proceedings of the 9th international conference on computer analysis of images and patterns, pp 317–324
2. Wark T, Sridharan S (1998) A syntactic approach to automatic lip feature extraction for speaker identification. In: Proceedings of IEEE international conference on acoustics, speech, and signal process, vol 6, pp 3693–3696
3. Goswami B, Christmas WJ, Kittler J (2006) Statistical estimators for use in automatic lip segmentation. In: Proceedings of the 3rd European conference on visual media production, pp 79–86
4. Cetingul HE, Yemez Y, Erzin E, Tekalp AM (2006) Discriminative analysis of lip motion features for speaker identification and speech-reading. IEEE Trans Image Process 15(10):2879–2891
5. Chetty G, Wagner M (2004) Automated lip feature extraction for liveness verification in audio-video authentication. In: Proceedings of the image and vision computing, pp 17–23
6. Dargham JA, Chekima A (2006) Lips detection in normalized color scheme. In: Proceedings of the 2nd IEEE international conference on information and communication technology, vol 1, pp 1546–1551
7. Eveno N, Caplier A, Coulon PY (2000) A new color transformation for lips segmentation. In: Proceedings of the 4th IEEE workshop on multimedia signal processing, pp 3–8
8. Lucey S, Sridharan S, Chandran V (2001) A suitability metric for mouth tracking through chromatic segmentation. In: Proceedings of the international conference on image processing, vol 3, pp 258–261
9. Liew AWC, Leung SH, Lau WH (2003) Segmentation of color images by spatial fuzzy clustering. IEEE Trans Fuzzy Syst 11(4):542–549
10. Wang SL, Lau WH, Liew AWC, Leung SH (2007) Robust lip region segmentation for lip images for complex background. Pattern Recogn 40(12):3481–3491

11. Nath R, Rahman FS, Nath S, Basak S, Audin SI, Fattah SA (2014) Lip contour extraction scheme using morphological reconstruction based segmentation. In: ICEEICT
12. Guan YP (2008) Automatic extraction of lips based on multi-scale wavelet edge detection. IET Comput Vis 2(1):23–33
13. Delmas P, Coulon YP, Fristot V (1999) Automatic snakes for robust lip boundaries extraction. In: Proceedings of the IEEE international conference on acoustics, speech, and signal processing, vol 6, pp 3069–3072
14. Xin L, Yiu-ming C, Hailin L (2010) A lip contour extraction method using localized active contour method with automatic parameter selection. In: Proceedings of the international conference on pattern recognition, pp 4332–4335
15. Vicent C, Ron K, Guillermo S (1997) Geodesic active contours. Int J Comput Vis 22(1):61–79
16. http://www.imm.dtu.dk/aam/datasets/datasets.html

Early Started Hybrid Denoising Technique for Medical Images

Khakon Das, Mausumi Maitra, Punit Sharma and Minakshi Banerjee

Abstract We propose an Early Started Hybrid Denoising Technique for brain images using modified Haar wavelet transform. To enhance the quality of the images, image fusion is used by combining two brain images obtained using positron emission tomography (PET) and computed tomography (CT) scan. Modified wavelet transform using lifting and in-place calculation has been proposed in the paper which has been shown efficient with respect to denoising parameters. The proposed algorithm is applied on the fused images, and the results show reduced noise parameters and enhanced denoised images. The peak signal-to-noise ratio (PSNR) and other statistical results like Universal Image Quality Index (UIQI) and Mutual Information (MI) are better for the fused images using modified Haar wavelet transform.

Keywords Image fusion · Denoising · Wavelet transform · In-place calculation Lifting

1 Introduction

Noise is a fundamental characteristic that is present in all images. Noise decreases the visibility level of some structures and entities, especially those that have quite low contrast. Operation on images at the lowest level of abstraction whose aim is

K. Das (✉) · M. Banerjee
RCC Institute of Information Technology, Kolkata 700015, India
e-mail: khokon.phd@gmail.com

M. Banerjee
e-mail: mbanerjee23@gmail.com

M. Maitra
Government College of Engineering & Ceramic Technology, Kolkata 700010, India
e-mail: mou1232005@yahoo.com

P. Sharma
Apollo Gleneagles Hospitals, Kolkata 700054, India
e-mail: dr_punitsharma@yahoo.com

S. Bhattacharyya et al. (eds.), *Recent Trends in Signal and Image Processing*,
Advances in Intelligent Systems and Computing 727,
https://doi.org/10.1007/978-981-10-8863-6_14

to enhance the image information without changing the basic image information content [1]. In medical imaging, the objective is to reduce the noise to a clinically acceptable level. In positron emission tomography (PET) imaging systems, the high-energy photons are emitted due to the annihilation of positrons and electrons, which are collected by the sensor. During the signal collection period, noise is also generated and added to the corresponding images. Denoising is the vital step for enhancing the quality of the medical images (PET, CT and MRI). Therefore, the type of the noise of the signal should be recognized, and its extent and statistical properties should be studied to design optimal denoising techniques for the recognized type of noise [2–5]. Details of image information like edge and curve are lost during denoising, which may be retrieved by using image edge preservation [6], image contrast enhancement [7, 8] and fusion of heterogeneous images like PET-CT [9–11].

Discrete wavelet transform (DWT) was first proposed by Mallat [12], which is widely used for denoising medical images. Several modifications of DWT have been made to reduce processing time and power as it involves huge computations [13–15]. A new concept of lifting scheme [16] was proposed by Sweldens in 1996 which is totally based on in-place calculation [17] and thereby eliminates the needs for auxiliary memory.

Depending on particular application, mother wavelet is selected. We have considered Haar wavelet as the mother wavelet for its simplicity and wide applications in medical image processing. In this work, modified wavelet transform using lifting and in-place calculation has been applied on brain CT and PET images separately for two patients, and then the denoised images are fused for obtaining better efficiency with respect to denoising parameters.

This paper is presented in the following way. Section 2 describes the proposed method. Different metrics for checking image quality have been described in Sect. 3. Results of the proposed work have been discussed in Sect. 4. Conclusion is presented in Sect. 5.

2 Proposed Work

We have developed a hybrid algorithm for denoising medical images using modified Harr wavelet and fusion of two CT and PET images. Here PET image gives us very good information about soft tissue, and CT image gives us the information about hard tissues or structures. “Early Started” refers that the algorithm starts performing wavelet transform as soon as information of the two pixels is received and algorithm stops at just after getting the last pixels’ information; so no need to wait for full image information at the time of algorithm begins. Our proposed algorithm is the modification of the basic Haar wavelet algorithm using lifting, in-place [17] and fusion techniques which have been described below:

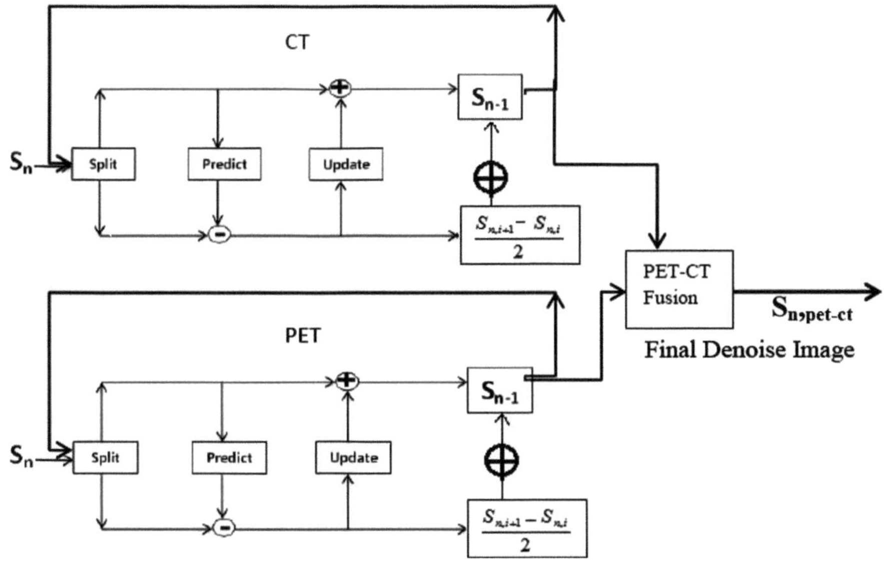

Fig. 1 Process diagram of the proposed scheme

Step 1: Split all the pixels into odd and even sets.

$$S\{Even\}\ \&\ S\{Odd\}\quad \text{Where } S_n = S_n\{Even\} \bigcup S_n\{Odd\} \tag{1.1}$$

["n" is a denoise level].

Step 2: Perform the predict step of lifting technique using in-place calculation for calculating the Haar's detail part.

$$S_{n,i} = S_{n,i+1} - S_{n,i} \tag{1.2}$$

["n" is a denoise level and "i" refers to pixel index].

Step 3: Perform update step of lifting technique using in-place calculation for calculating the Haar's approximate part of the image.

$$S_{n,i} = S_{n,i} + \left(\frac{S_{n,i+1} - S_{n,i}}{2}\right) \tag{1.3}$$

Step 4: Lifting with fusion

- 4.1: Resize the PET image by a factor K. [K is dependent on PET-CT scanner, here K = 1.5]
- 4.2: Make a square window of resized PET image and map the centre of the window to the seed point of the CT image for the same slice of both the images (Fig. 1).

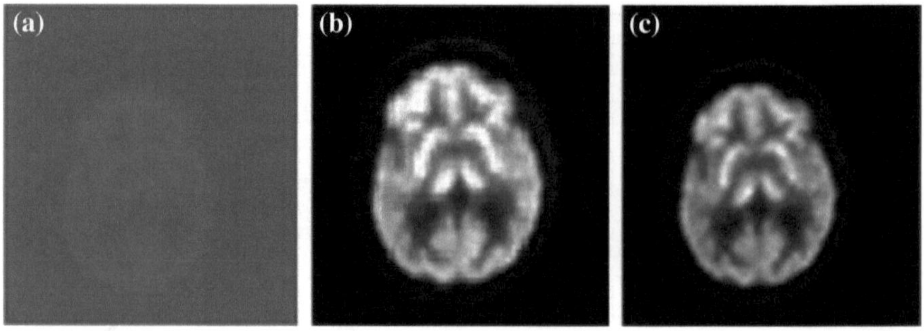

Fig. 2 Input and output images of DWT2 and the proposed method (using PET image for Patient 1)

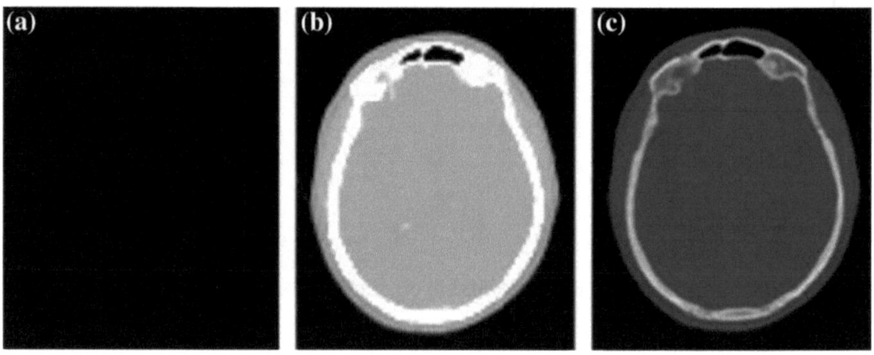

Fig. 3 Input and output images of DWT2 and the proposed method (using CT image of Patient 1)

4.3:

$$S_{n,PET-CT}(index) = \frac{S_{n,PET}(index) + S_{n,CT}(index)}{2} \quad (1.4)$$

[$S_{n,PET-CT}$ is the fused image].

Figure 2a shows the input PET image of Patient 1, and Fig. 2b shows the corresponding denoised image using DWT2 while Fig. 2c shows the denoised PET image using the proposed method.

Figure 3a shows the CT image for the same patient, and Fig. 3b shows the corresponding denoised image using DWT2 while Fig. 3c shows the denoised CT image using the proposed method.

Output image after fusion: Fig. 4a shows the input PET image, Fig. 4b shows the input CT image, and Fig. 4c shows the fused PET-CT image before denoising.

Figure 5a, b shows the output PET images for Patient 1 after Level 1 and Level 2 denoising using our proposed method while the same have been shown in Fig. 5c, d for Patient 2.

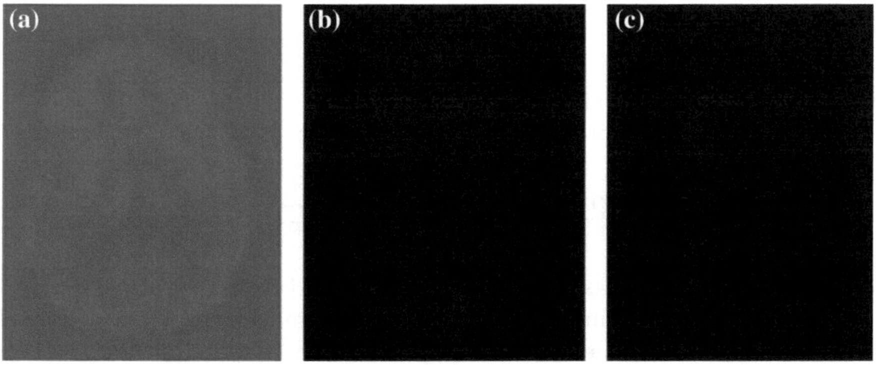

Fig. 4 Output images after fusion (before denoising)

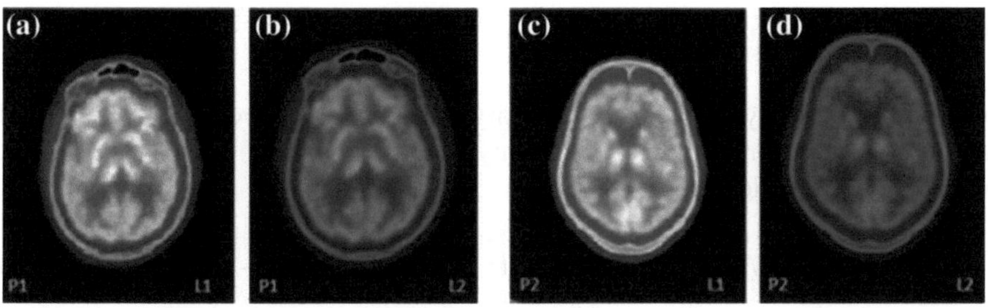

Fig. 5 Output images after fusion (after denoising)

3 Parameters for Measurement of Image Quality

3.1 Error Metrics

The mean square error (MSE) and the peak signal-to-noise ratio (PSNR) are the two important error metrics for comparing image denoising techniques [18], which are given by the mathematical formulae below

$$MSE = \frac{1}{MN}\sum_{y=1}^{M}\sum_{x=1}^{N}\left[I(x, y) - I'(x, y)\right]^2 \tag{1.5}$$

$$PSNR = 20 * \log_{10}\left(\frac{255}{\sqrt{MSE}}\right) \tag{1.6}$$

where I(x, y) and the I′(x, y) are the original and the denoised images of dimension (M × N).

3.2 Universal Image Quality Index (UIQI)

It indicates the image or video quality distortion. For two images x[i, j], y[i, j], the UIQI Q is calculated as a product of three components:

$$Q = \frac{\sigma_{xy}}{\sigma_x \sigma_y} \cdot \frac{2\overline{x}\overline{y}}{(\overline{x})^2 + (\overline{y})^2} \cdot \frac{2\sigma_x \sigma_y}{\sigma_x^2 + \sigma_y^2} \quad (1.7)$$

where the first component is the correlation coefficient between x and y, the second component measures the similarity between the mean luminance between x and y, and the third component measures the similarity of the contrast of the two images [19].

3.3 Mutual Information (MI)

It gives the similarity of the image intensity distributions of two images. Let X and Y be two random variables with probability distributions pX(x) and pY(y) and joint probability distribution pXY(x, y) [20].

$$I_{XY}(x, y) = \sum_{a,b} P_{XY}(x, y) \log \frac{P_{XY}(x, y)}{P_X(x) P_Y(y)} \quad (1.8)$$

Considering two input images A, B and a fused image F, we can calculate the amount of information that F contains about A and B according to above equation

$$I_{FX}(f, x) = \sum_{a,b} P_{FX}(f, x) \log \frac{P_{FX}(f, x)}{P_F(f) P_X(x)} \quad (1.9)$$

$$I_{FY}(f, y) = \sum_{a,b} P_{FY}(f, y) \log \frac{P_{FY}(f, y)}{P_F(f) P_Y(y)} \quad (1.10)$$

Thus, the Mutual Information is given by

$$M_F^{XY} = I_{FX}(f, x) + I_{FY}(f, y) \quad (1.11)$$

4 Results and Discussion

The proposed algorithm was applied to original PET and CT brain images of two patients as obtained from Apollo Gleneagles Hospitals, Kolkata. From Table 1, we can say that MSE and PSNR of PET and CT images for both the patients are

Table 1 MSE, PSNR and UIQI using DWT2 and proposed method for Patient 1 and Patient 2

Algorithm		CT image			PET image		
		MSE	PSNR (dB)	UIQI	MSE	PSNR (dB)	UIQI
Patient 1							
DWT2	Input image	68108.84842	±0.33447	0.38655	537114.17169	±18.27173	0.49115
	Level 1	12.57570	+74.33896	0.24657	43.96118	+63.46821	0.76160
	Level 2	13.12524	+73.96745	0.40996	50.08105	+62.33613	0.95319
	Level 3	14.84521	+72.89787	0.65841	46.19531	+63.03764	0.99300
Modified DWT	Input image	68108.08267	±0.33437	0.38461	537052.73871	±18.27074	0.49312
	Level 1	10.92262	+75.56307	0.34129	42.33423	+63.79577	0.69068
	Level 2	0.87052	+97.53401	0.80136	3.88770	+84.53575	0.75694
	Level 3	0.07698	+118.60197	0.91995	0.29688	+106.87813	0.74180
Patient 2							
DWT2	Input image	71598.58635	±0.76849	0.37083	202780.91772	±9.81094	0.49869
	Level 1	12.57246	+74.34119	0.23655	43.79785	+63.50054	0.75050
	Level 2	12.72437	+74.23688	0.39101	47.58691	+62.77985	0.94070
	Level 3	15.23706	+72.67157	0.63612	64.67578	+60.11476	0.98659
Modified DWT	Input image	71601.99099	±0.76890	0.36938	202646.73016	±9.80519	0.49923
	Level 1	10.71794	+75.72737	0.33419	29.08185	+67.05716	0.67821
	Level 2	0.88059	+97.43410	0.79993	2.62781	+87.93773	0.75090
	Level 3	0.07565	+118.75394	0.92358	0.24139	+108.67507	0.73512

Table 2 MSE, PSNR, MI and UIQI of fused images

Algorithm			PET-CT fused image			
			MSE	PSNR (dB)	MI	UIQI
Modified DWT	Patient 1	Input image	57854.90155	+1.08280	4.39480	0.40392
		Level 1	10.17807	+76.17629	4.36313	0.37228
		Level 2	0.80376	+98.22711	4.37061	0.80639
		Level 3	0.05933	+120.86356	4.37566	0.93035
	Patient 2	Input image	38049.34261	+4.72266	4.30599	0.38623
		Level 1	9.10414	+77.14482	4.28492	0.36254
		Level 2	0.70793	+99.32977	4.33545	0.80434
		Level 3	0.05832	+121.01282	4.32020	0.93151

better for our proposed method, while in case of UIQI our proposed method is little behind from DWT2. But proposed method on fused images shows improvement in the values of MSE, PSNR and UIQI as presented in Table 2. Figures 6, 7 and 8 show that DWT2 has poor MSE and PSNR compared to our proposed method. When the denoised images are fused, more enhanced images are obtained with respect to edges and corners and reduce localized errors. The "update part" helps to remove noise by subtracting the half of difference value from the image where difference is a

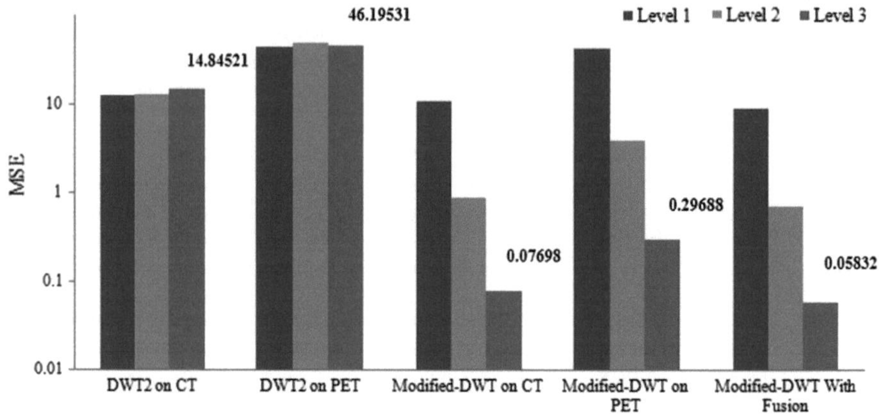

Fig. 6 Comparison of MSE values of DWT2, modified DWT

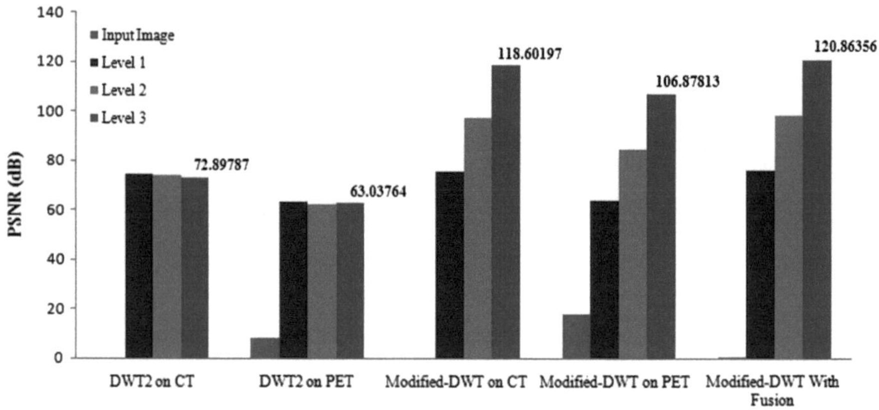

Fig. 7 Comparison of PSNR values of DWT2, modified DWT

high-frequency part of the image. In fusion part of our algorithm, we have used seed point and window-based techniques to fuse two heterogonous images like PET and CT. The PET-CT fused image is more informative than PET image and CT image individually. It also provides very good MI and UIQI. Mutual Information (MI) [20] value refers to the amount of information hold by the PET-CT fused image about PET and CT Images. Therefore, the structural information is copied from CT image to PET-CT fused image as it is clearly seen in Fig. 5.

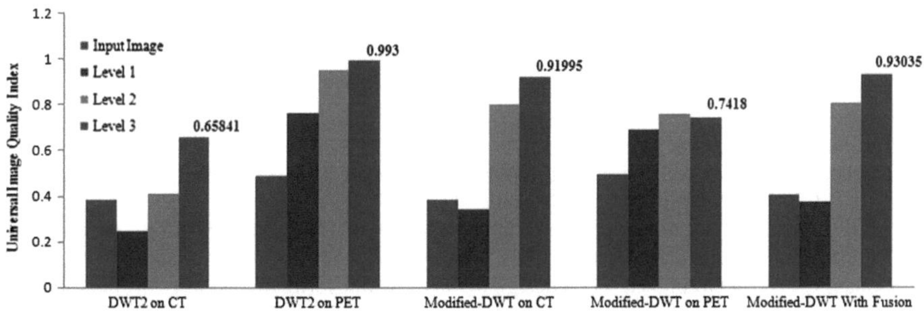

Fig. 8 Comparison of UIQI values of DWT2, modified DWT

5 Conclusion

In this paper, we have tried to focus on hybrid techniques for medical image denoising using heterogeneous image fusion and modified Haar wavelet transform. PET and CT brain images are considered for two patients for our studies. From the various graphs and denoising parameters obtained by the proposed method, it can be concluded that the overall performance of this method is good in terms of PSNR, MSE, MI and UIQI.

Acknowledgements We acknowledge BRNS, DAE, Govt. of India, for sponsoring fellowship to Mr. K Das, JRF in The Project "Embedding Segmentation of PET Images for Medical Applications in Brain Disorders".
We have obtained all ethical approvals from an appropriate ethical committee. The consents of the participants have also been taken to use the data in this research. Neither the editors nor the publisher will be responsible for any misuse or misinterpretation of the data.

References

1. Sonka M, Hlavac V, Boyle R (1993) Image pre-processing. In: Chapter—Image processing, analysis and machine vision. Springer, US, pp 56–111, ISBN: 978-0-412-45570-4
2. Yu S, Muhammed HH (2016) Noise type evaluation in positron emission tomography images. In: 1st international conference on biomedical engineering (IBIOMED), Stockholm, Sweden
3. Unser M, Aldroubi A (1996) A review of wavelets in biomedical applications. Proc IEEE 84(4)
4. Nowak RD (1999) Wavelet-based Rician noise removal for magnetic resonance imaging. IEEE Trans Image Process 8(10)
5. Portilla J, Strela V, Wainwright MJ, Simoncelli EP (2003) Image denoising using scale mixtures of Gaussians in the wavelet domain. IEEE Trans Image Process 12(11)
6. Verma K, Singh BK, Thoke AS (2015) An enhancement in adaptive median filter for edge preservation. Procedia Comput Sci 48:29–36
7. Singh S, Pal P (2016) Contrast enhancement of medical images: a review. IJRDO-J Health Sci Nurs 1(4), Paper-4
8. Nancy, Kaur S Er (2013) Image enhancement techniques: a selected review. IOSR J Comput Eng (IOSR-JCE), 9(6):84–88
9. Chandana M, Amutha S, Kumar A (2011) A hybrid multi-focus medical image fusion based on wavelet transform. Int J Res Rev Comput Sci 2:1187–1192

10. Agarwal J, Bedi SS (2015) Implementation of hybrid image fusion technique for feature enhancement in medical diagnosis. In: Human-centric computing and information sciences. SpringerOpen Journals
11. Tawade L, Aboobacker AB, Ghante F (2014) Image fusion based on wavelet transforms. Int J Bio-Sci Bio-Technol 6(3):149–162
12. Mallat SG (1989) A theory for multi-resolution signal decomposition: the wavelet representation. IEEE Trans Pattern Anal Mach Intell 11(7):674–93
13. Chilo J, Lindblad T (2008) Hardware implementation of 1D wavelet transform on an FPGA for infrasound signal classification. IEEE Trans Nucl Sci 55(1):9–13
14. Souani C (2004) Optimized VLSI design of wavelet transform architecture. In: International conference on ICM proceedings microelectronics, Tunis, Tunesia, 6–8 Dec 2004, pp 558–563
15. Motra AS, Bora PK, Chakrabarti I (2003) Efficient hardware implementation of DWT and IDWT. In: Conference on convergent technologies for Asia-Pacific region, Bangalore, India, vol 1, 15–17 Oct 2003, pp 95–99
16. Daubechies I, Sweldens W (1998) Factoring wavelet transform into lifting steps. J Fourier Anal Appl 4(3):247–69
17. Koichi K, Koichi N, Shigeru T (2004) FPGA-based lifting wavelet processor for real-time signal detection. Signal Process (Elsevier) 84(10):1931–40
18. Huynh-Thu Q, Ghanbari M (2008) Scope of validity of PSNR in image/video quality assessment. Electr Lett 44(13)
19. Wang Z, Bovik AC (2002) Universal image quality index. IEEE Signal Process Lett 9(3)
20. Pluim JPW, Maintz JBA, Viergever MA (2003) Mutual-information-based registration of medical images: a survey. IEEE Trans Med Imaging 22(8)

Intelligent Tutoring by Diagram Recognition

A. Mondal, Anirban Mukherjee and U. Garain

Abstract It is a proven psychological fact that every student has his own different way of learning. Therefore, effective teaching should be student specific, not a general one. Our aim is to design a computer-based, adaptive learning system that assesses a child's understanding of a given problem and guides most effectively to correctly understand the problem and the related concept in turn. As a case study, problems of elementary geometry of primary school level are considered. The system would allow a student to follow a learning path that is based on his ability of understanding and the pace he is comfortable with. It would help improve the confidence and skill of the child in geometry by presenting geometrical figures of increasing difficulty level.

Keywords Pattern recognition · Geometry drawing · Computer-based teaching–learning · Artificial intelligence

1 Introduction

By-hand constructions and drawing using ruler and compass play an integral role in constructive geometry; such tasks are included in all textbooks for schools. Many teachers believe that compass and straight-edge constructions on a sheet of paper are the means leading to a proper understanding of geometry.

A. Mondal (✉)
Department of Computer Application, RCC Institute of information Technology, Kolkata, India
e-mail: arindammond@gmail.com

A. Mukherjee
Department of Engineering Science & Management, RCC Institute of Information Technology, Kolkata, West Bengal, India
e-mail: anirbanm.rcciit@gmail.com

U. Garain
CVPR Unit, Indian Statistical Institute, Kolkata, India
e-mail: utpal.garain@gmail.com

S. Bhattacharyya et al. (eds.), *Recent Trends in Signal and Image Processing*,
Advances in Intelligent Systems and Computing 727,
https://doi.org/10.1007/978-981-10-8863-6_15

With computers being increasingly used as a teaching–learning tool, a few software have been developed that replaces the use of ruler and compass to teach and learn geometry construction at all levels. Students at primary level can learn geometry through some basic shapes like lines, square, rectangle, triangle using dynamic mathematics software like GeoGebra [1]. There are other geometry drawing software like Geometria, Sketchometry, Sketchpad, Geomspace [2]. All of these software provides an interface for drawing the geometry pattern. LIM-G [3] is another knowledge-based prototype that understands and draws elementary school-level geometry diagrams.

A new study [4] suggests that students who used dynamic geometry software were more successful in discovering new mathematical ideas than when they used static, paper-based construction. Incorporating technology in high school-level geometry classes not only makes the teaching of concepts such as congruency easier, it also empowers students to discover other geometrical relationships they would not ordinarily uncover when traditional methods of instruction and construction were used.

Though such software seems to be useful in teaching-learning of geometry, those are not student specific. Herein comes a question of how a pre-programmed learning exercise can be equally useful to students with varying level of understanding and varying need for teaching input or guided path for correct understanding of a given geometry concept. This is probably the limitation of the geometry software available in the market and also the reason why many teachers prefer the traditional paper and pencil or chalk and board method of teaching-learning geometry. A teacher, while physically interacting with a group of students, intuitively realizes the difficulty of understanding a certain concept by a student and accordingly changes the teaching content and context. Often the teacher has to break up a geometry problem or figure into its constituent parts or conceptual components or even demonstrate simpler problem to help in the understanding process of the student. On the contrary, a student with a better level of understanding is usually given harder problems to solve. The pace at which the learning inputs are presented often needs to be varied to suit the comfort or ability of a student to accept the input. The teacher has his own method based on his experience and knowledge to test a students' understanding and the teaching pace with which the student is comfortable. When the teacher to student ratio is more than 1:15 or so, it becomes very difficult for a teacher to individually guide a student and therein emerges the requirement of automated intelligent tutoring. To replace the teacher's role with intelligent tutoring software is a challenging task and requires the application of artificial intelligence (AI) to take the human factors into consideration to some reasonable extent.

In this paper, a feasible automated approach has been described that can handle different needs of learning or level of understanding of a student as far as elementary geometry figures are concerned. Besides presenting learning inputs (figures of increasing conceptual difficulty) in a step-by-step manner, the proposed system would test the students' understanding of geometry concepts in order of increasing difficulty. Based on the performance (correctness of figure drawn) by a particular student at each difficulty level, he would be tested with figures of higher or lower difficulty level. Thus, there will be dynamic assessment of students' performance or understanding and also re-learning of related basics to help the students achieve a complete understanding of elementary geometry through varying time and steps.

The proposed system can also help in examining those children who suffer from visual perceptual disease. This type of disorder affects the understanding of information that a person sees and/or affects his ability to draw or copy. The system as described here can measure and record incorrect figure drawing by a child which might be helpful to identify how much visual perceptual disorder the child has.

2 Methodology

The first step in automated intelligent tutoring of geometry at primary school level is to demonstrate the fundamental geometrical concepts and entities in a step-by-step manner. Starting from elementary concepts like horizontal line, vertical line, slant line, perpendicular lines, parallel lines, intersecting lines, simple to complex entities like rectangle, square, rhombus, parallelogram, pentagon, hexagon, triangle, and their types and then such entities with bisectors, diagonals, medians, etc., are demonstrated in the lesson mode. The demonstration can be repeated from any level at any point of lesson mode run-time as may be desired by a student. In the test mode, which forms the most important part of the proposed system, the students' understanding of lessons will be tested. The student would be asked to draw geometrical figures chosen randomly from the set of figures pertaining to each difficulty level. The correctness of the geometrical figure drawn by a student in the computer can be evaluated by the proposed system following the method proposed by Mukherjee et al. in their earlier work [5, 6]. Upon identification of incorrectness, the student can be trained up with easier figure or more fundamental concept. Thus, to make the tutoring of geometry really student specific, the difficulty level of the preselected figures or concepts are assigned first, then those figures are categorized according to difficulty level, and are then connected in a hierarchy of related figures of increasing difficulty level. This yields a figure database covering the entire learning content (for the primary school-level geometry as considered in this case study). The figure database containing the figure description as well as the related geometrical properties of each figure forms the core of the proposed intelligent system. The interface of the system is designed to display text of the geometrical figure description and provide grid for students to draw straight line figures. Let us consider a student who is asked to draw a geometric entity, e.g., 'Draw a parallelogram.' When the student draws a figure using the grid, the system automatically checks whether the figure drawn is correct, i.e., a true parallelogram or not (by matching with the properties of parallelogram already stored in the figure database). If the figure is correct, then it asks to draw a figure of next higher level of difficulty, e.g., 'Draw a parallelogram and one bisector of opposite angles.' If the first attempt is partially correct, then it asks another question of same level of difficulty, e.g., 'Draw a square.' But if it is totally incorrect (matchscore is below a threshold level), then it asks a lower-level question, e.g., 'Draw two parallel lines.' From stored definition of figures, the correct diagram for every question can be

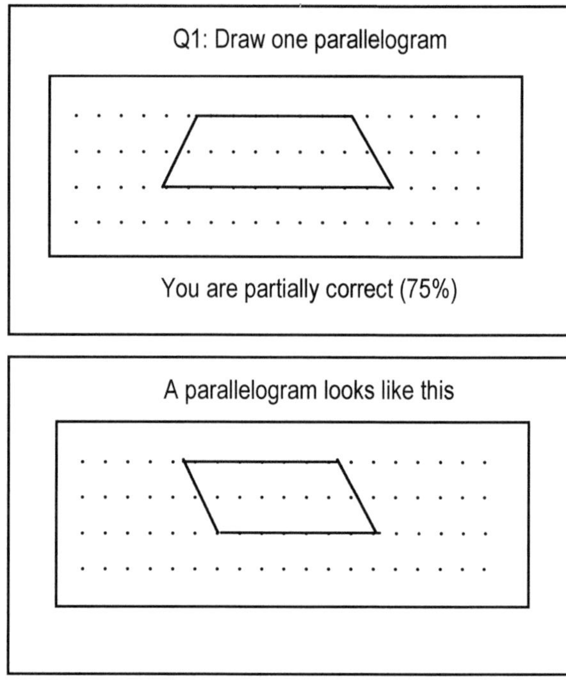

Fig. 1 Interface for geometrical figure drawing and display

optionally displayed for the student for him to learn the error. Like this, the progress of a student can be tracked and student's performance or score can be calculated.

3 Input

Let us consider a student who accesses the proposed tool and starts with a simple geometrical figure. The interface provides a grid of dots on the screen. When the student clicks with mouse near one of the grid points, it is selected and a line starting from that point is shown as dragline (rubber band line). When he again clicks near another grid point, then a firm line is drawn up to it. Thus, the student can draw a figure with a set of straight line segments. Once the figure is completely drawn, the student needs to click 'submit' button for the program to evaluate the correctness of the figure and return with the feedback 'correct,' 'partially correct,' and 'incorrect' with scores as 100, 50–99, and 0–49, respectively (Fig. 1).

As the grid points are predefined, so the program can extract the coordinates of the vertices or line endpoints of the figure drawn by the student in the order the points are connected by line segments. The coordinate points of the correct geometrical figure to be displayed when the student draws incorrectly are preselected and stored in the figure database.

4 Assessment

For checking the extent of matching of the figure drawn by student and the figure definition stored in the machine, the method proposed by Mukherjee et al. [6] can be adopted.

Once the coordinates of the figure drawn by the students are extracted in the order they were connected, the adjacency matrix is automatically generated. Subsequently, the system calculates the length matrix and slope matrix for the edges of the figure drawn. From the slope matrix, the system computes the smaller (acute) angle between every pair of edges connected at any point. Then, it searches for angles of same values and forms different sets of same angles (if more than one). Similarly, sets are found for same-length edges of the figure drawn. Values for the same parameters corresponding to the correct figure (whose coordinates are preselected) are stored in the figure database.

Next, the system evaluates the geometrical similarity of the student-drawn figure and machine-stored figure by comparing the values of the following parameters.

(i) Adjacency matrix
(ii) Total number of points
(iii) Total number of edges
(iv) Total number of angles between edges
(v) Sets of same-length edges
(vi) Sets of same angle between edges

If at first comparison, the adjacency matrices (of same order) of two figures do not match element by element, then it is checked whether they match if one of the matrices is read in reverse row and/or reverse column direction.

With regard to parameters (ii) to (vi), the scores are simply calculated as the ratio of the parameter value (say x) for the student-drawn figure to that (say y) of the machine-stored figure provided $x \leq y$. The score x/y should be equal to 1 for all the parameters if the figure is 100% correct. If $x > y$ for any parameter, then the corresponding score is calculated as $1 - \{(x - y)/y\}$, thus assigning penalty for $(x - y)$ additional elements generated wrongly against expected y elements. There will be multiple values of parameter for (v) and (vi).

The cumulative score of the student-drawn figure considering all the parameters and their multiple values is calculated. The sum of the maximum possible scores against each of these parameter values is stored in the database. The maximum possible score corresponds to a 100% correctly drawn figure. The final output of evaluation is the percentage of cumulative score of the student-drawn figure, and this is calculated as follows:

$$\frac{\sum \text{scores\_of\_parameters (i) to (vi)}}{\sum \text{max\_scores\_of\_parameters (i) to (vi)}} \times 100$$

Table 1 Score sheet for example case

	Adjacency matrix	Points	Edges	Angle	Same length edge	Same angle
y value	4 × 4	4	4	4	2, 2	4
x value	4 × 4	4	4	4	0	2, 2
Score	1	1	1	1	0	0.5
Max_score	1	1	1	1	1	1

5 Experimental Result

For the example cited in Sect. 3, processing is done as shown in Sect. 4 and the comparison summary with respect to the values of six parameters is tabulated in Table 1.

The above summary is computed based on the intermediate output given by the system, i.e., the slope matrix, edge length matrix, same length edge sets, and same angle sets, as depicted in Table 2.

In this case, the cumulative score of the student-drawn figure is 4.5, while the maximum possible score is 6. This implies that the match percentage is [(4.5/6) * 100], i.e., 75%.

So as evaluated by the machine, the student is partially correct (the score is within 50–99%). Hence, the student gets a new question (figure) of same level. The system finds such figure from the database of figures stored in a hierarchy of difficulty level. A pictorial representation of a part of such figure database is shown in Fig. 2. It is evident from Fig. 2 that in the example case described above, the student will be given a figure of level 2, i.e., to draw a rectangle or square or rhombus.

6 Conclusion

This paper proposes a feasible implementation of a student-specific automated system of tutoring of geometrical concepts. Elementary geometrical figures for the primary school students are taken as a case study; a drawing grid is presented to in the machine for the students to draw figures of a geometrical concept or entity. The coordinates and related parameter values of all the correct figures corresponding to each geometrical concept or entity are stored previously. The parameter values can be used as a reference to compare the correctness of a figure actually drawn by a student. The correctness can be assumed as his extent of understanding a concept. Though this method of automatically evaluating the correctness of a student-drawn digital geometrical figure is not foolproof, still this is the only available method of recognition of geometrical line diagrams. Evaluating complex diagrams with multiple entities including circle or curved lines still remains a challenge. The concept

Table 2 Analysis of example case

	Correct figure	Student drawing
Figure	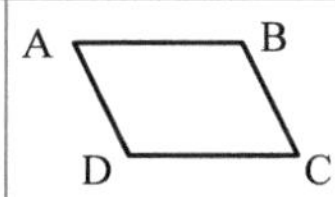	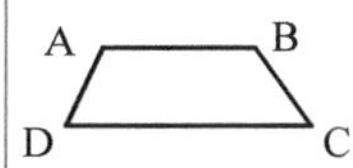
Coordinate	A (10, 10), B (10, 40), C (30, 50), D(30, 20)	A (10, 20), B (10, 40), C (30, 50), D (30, 5)
Adjacency matrix	0 1 0 1 1 0 1 0 0 1 0 1 1 0 1 0	0 1 0 1 1 0 1 0 0 1 0 1 1 0 1 0
Total no. of points	4	4
Edge slope matrix	999 0 999 0.5 0 999 0.5 999 999 0.5 999 0 0.5 999 0 999	999 0 999 −0.75 0 999 0.5 999 999 0.5 999 0 −0.75 999 0 999
Edge length matrix	0 30 0 22.36 30 0 22.36 0 0 22.36 0 30 22.36 0 30 0	0 20 0 2.91 20 0 22.36 0 0 22.36 0 45 22.91 0 45 0
Total no. of edges	4	4
Same length edge sets	30 ->2 22.36 ->2	0
Total no. of angles	4	4
Same angle sets	0.5 = 4	0.75 = 2 0.5 = 2

of attaching difficulty level to the individual drawing exercises and automatically guiding a student back and forth between the easier and harder figures according to his performance emulates the act of a teacher. Thus, the same system can be used to provide uniform learning input to all the students, while at the same time, it can be used to test a students' understanding of the concepts learned and accordingly help him re-learn the basics as much as required to consolidate his understanding.

The proposed system can be further extended to enhance the difficulty level by giving only blank space for drawing (removing the grids) similar to white sheet used by higher class students. In that case, recording the pen-up, pen-down events will enable to capture the key points or vertices drawn by the students. At high school level, the variation of geometry problems is so much that it is not possible to store the figure or its parameter values corresponding to each such problem. In that case, the text-to-drawing tool based on GeometryNet (an earlier work by Mukherjee et al.

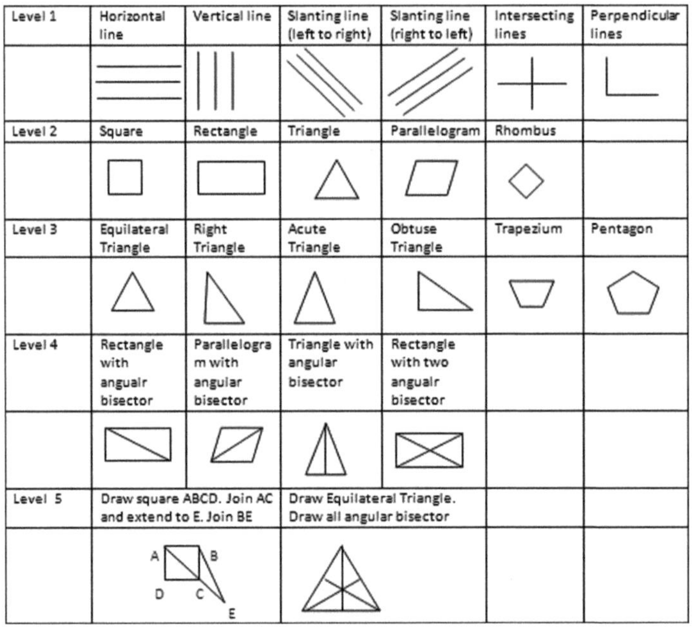

Fig. 2 Pictorial representation of figure database (partial)

[7, 8]) will play a significant role in dynamically generating coordinates to represent the figure correctly and generate comparable parameter values from the problem text itself. Thus, the proposed system can go a long way in developing a comprehensive automated solution for intelligent tutoring and assessment of school-level geometry.

References

1. https://www.geogebra.org/
2. https://en.wikipedia.org/wiki/List_of_interactive_geometry_software
3. Wong WK, Hsu SC, Wu SH, Hsu WL (2007) LIM-G: learner-initiating instruction model based on cognitive knowledge for geometry word problem comprehension. Comput Educ J 48(4):582–601
4. Skultety L, Gonzalez G, Vargas G (2017) Using technology to support teachers' lesson adaptations during lesson study. J Technol Teacher Educ 25(2)
5. Mukherjee A, Garain U, Biswas A (2013) Evaluation of the graphical representation for text-to-graphic conversion systems. In Proceedings of 10th IAPR international workshop on graphics recognition—GREC 2013, Bethlehem PA, USA
6. Mukherjee A, Garain U, Biswas A (2014) Evaluation of diagrams produced by text-to-graphic conversion systems, graphics recognition. In: Lamiroy B, Ogier J-M (eds) Current trends and challenges, GREC 2013. LNCS 8746. Springer, Chapter 20, pp 252–265

7. Mukherjee A, Garain U (2011) Intelligent tutoring of school level geometry using automatic text to diagram conversion utility. In: Proceedings of 2nd international conference on emerging applications of information technology (EAIT), Kolkata, India. IEEE, pp 45–48
8. Mukherjee A, Garain U, Nasipuri M (2007) On construction of a GeometryNet. In: Proceedings of the IASTED international conference on artificial intelligence and applications, Calgary, Canada. ACTA Press, pp 530–536

Color MRI Image Segmentation Using Quantum-Inspired Modified Genetic Algorithm-Based FCM

Sunanda Das, Sourav De, Siddhartha Bhattacharyya and Aboul Ella Hassanien

Abstract A quantum counterpart of classical modified genetic algorithm-based FCM is presented in this article for color MRI image segmentation. Though classical modified GA enhances the global search optimality of conventional GA but to speed up and make it more optimal and cost effective, some quantum computing phenomena like qubit, superposition, entanglement, quantum gate are incorporated here. To achieve the final segmented output, the class levels generated by quantum-inspired modified genetic algorithm are now fed to conventional FCM to overcome the early convergence to local minima problem of FCM. A performance comparison is delineated between quantum-inspired modified GA-based FCM, quantum-inspired GA-based FCM and classical modified GA-based FCM based on two color MRI images.

Keywords MRI image · Image segmentation · Modified genetic algorithm Quantum-inspired soft computing · FCM

S. Das (✉)
Department of Computer Science and Engineering, University Institute of Technology, Burdwan, West Bengal, India
e-mail: sunanda0301@gmail.com

S. De
Department of Computer Science and Engineering, CoochBehar Government Engineering College, CoochBehar, West Bengal, India
e-mail: dr.sourav.de79@gmail.com

S. Bhattacharyya
Department of Computer Application, RCC Institute of Information Technology, Kolkata, West Bengal, India
e-mail: dr.siddhartha.bhattacharyya@gmail.com

A. E. Hassanien
Information Technology Department, Cairo University, Cairo, Egypt
e-mail: aboitcairo@gmail.com

S. Bhattacharyya et al. (eds.), *Recent Trends in Signal and Image Processing*, Advances in Intelligent Systems and Computing 727,
https://doi.org/10.1007/978-981-10-8863-6_16

1 Introduction

Medical image processing is enriched every day by different innovative ideas of researchers. Among them medical image segmentation is widely cultured as it is considered as the most challenging and complex work due to some intrinsic properties of medical images. MRI imaging is the process where different anatomical structures of body are portrayed into an image with the use of radioactive magnetic field. An MRI image mainly contains white matter, gray matter, cerebrospinal fluid, and different vasive and non-invasive cells. A color MRI image holds much more information than grayscale image; so a proper segmentation technique is essential to recognize the affected portion accurately. MRI image segmentation is mainly used to identify tumors, classification of tissues and blood cells, multimodal registration, etc.

Segmentation is the process where pixels having similar properties make groups based on some features like intensity, color, texture, homogeneity. A color image is actually a combination of three color components viz. Red, Green, and Blue and their admixtures. Different classical, soft computing, and hybrid systems are popularly used to segment color images. Fuzzy C-Means clustering [1], a soft clustering technique, is widely used for MRI image segmentation. This algorithm follows the rule of fuzzy set theory. Though it is easy to implement and takes less time than any evolutionary algorithm, it suffers with some disadvantages like it has the possibility to stuck to local minima point easily if the initial cluster centers are not defined properly, which leads to inappropriate segmentation results. Also it does not consider the spatial information thus it is highly sensitive to noise. Ahmed et al. [2], Yang et al. [3], Adhikari et al. [4] introduced different modified FCM algorithms where spatial information is taken into consideration. Chaabane et al. [5] proposed a modified FCM algorithm where statistical features are merged with FCM to segment color images. Another histogram-based FCM method is introduced where histogram thresholding method is used to generate class levels for color images that are fed to FCM for final segmentation purpose [6]. A Fuzzy C-regression model is proposed by Chen and Ludwig [7] for color image segmentation where spatial information is included, and it is also applicable for hyperplane clusters.

Different evolutionary algorithms combined with the soft computing techniques are introduced in different research papers. Multilevel self-organizing neural network (MLSONN) in a combination with parallel optimized multilevel sigmoidal (ParaOptiMUSIG) activation function is used for color image segmentation [8, 9]. Here, GA is used to generate the ParaOptiMUSIG activation function. GA in combination with the weighted undirected graph is employed to segment the color images [10]. With the use of both supervised and unsupervised techniques, a color image segmentation method is evolved [11]. Here, SOM and simulated annealing are used for color reduction and clustering purpose. Afterward, supervised learning is applied to get the final segmented output. GA and PSO are also merged with FCM to overcome its inefficiencies. Castillo et al. [12] presented an optimized FCM algorithm with the use of GA and PSO. Mukhopadhyay and Maulik [13] suggested a multiobjective real-coded genetic fuzzy clustering algorithm for the segmentation of multispectral MR images of the human brain. A new method is approached in [14] for color image segmentation where a modified version of GA is combined with FCM to get more optimal results. Modified GA enhances efficacy of conventional GA by modifying its population initialization and crossover part. The optimized class levels generated by modified GA (MfGA) are employed to FCM as initial cluster centers to overcome the problem of stucking to local minima point.

All the hybrid systems are more efficient than the conventional methods and they all produce convenient results, but they take a huge computational time. Computations are performed with the help of different electronic circuits. High usage of electronic circuits makes the system costly; so a high computational time makes a system more costly. To bring down with this problem and also to find much more optimized accurate and convenient result, some quantum computing phenomena like entanglement, orthogonality, interference, quantum logic gates are incorporated to classical systems. Talbi et al. [15] introduced an quantum version of GA which is used to solve the traveling salesman problem within a minimum computational time. In [16], Talbi and Draa introduced a hybrid concept to solve the real parameter optimization problem with the help of quantum-inspired evolutionary algorithm. In this method, a proper sampling of search space is used for QEA to accommodate continuous optimization without losing superposition states. A quantum-based PSO algorithm is used for color image segmentation [17]. To maintain superposition states and inherent parallelism, Caraiman and Manta [18] implement a quantum oracle circuit to produce thresholding function which is used for threshold-based image segmentation. Dey et al. [19] incorporated quantum computing phenomena to particle swarm algorithm and differential evolutionary algorithm for multilevel thresholding of color images.

In this article, we provide a quantum counterpart of modified genetic algorithm-based FCM clustering method [14] to make this system less cost effective, less time consuming, and more optimized and realistic. It is already seen that this MfGA-based FCM outperforms GA-based FCM and the conventional FCM. A performance comparison of quantum-inspired MfGA-based FCM, quantum-inspired GA-based FCM, and classical MfGA-based FCM for two color MRI images based on standard statistical measure $F(I)$ [20] and $Q(I)$ [21] are reported in this article.

2 Quantum Computing

Quantum computing concept is based on the principle of quantum mechanics. Quantum systems are used nowadays as it takes less computational time than the classical systems. High computational time indicates usage of large amount of electronic circuits which makes classical systems more costly. To overcome this problem and also make classical systems more accurate, quantum computing concept is evolved by American computer engineer Howard Aiken in 1947. In quantum systems, the computational space increases exponentially with the size of the system which possess exponential parallelism thus the quantum systems are faster than its classical counterpart.

Quantum system is described using a wave function ψ which exists in Hilbert space. The smallest unit of quantum system is called **Qubit**. In classical systems, a bit can be represented by either 0 or 1, but in quantum computing a qubit is represented by 0 or 1 or any linear superposition of both states. If a sphere is taken as consideration, then it can be said that in classical computing a bit can reside only at two poles but in quantum system it can be any point of the sphere. A qubit can be represented as

$$|\psi\rangle = \alpha|0\rangle + \beta|1\rangle \tag{1}$$

where $|0\rangle$ and $|1\rangle$ represent the classical bit value 0 and 1 and α and β are complex numbers having the property of

$$\alpha^2 + \beta^2 = 1 \tag{2}$$

This property is called as **Orthogonality**. Another quantum property **Entanglement** can be stated as where states are highly correlated to each other that they can not be separated independently. As qubit represents a linear superposition state of 0 and 1 so at a time it holds multiple states. Theoretically it is defined as a tensor product between two states, written as $|\vartheta_1\rangle \bigotimes |\vartheta_2\rangle$.

Quantum measurement is the probabilistic measurement where entangled states are converted to its corresponding state, i.e., the wave function ψ is converted into single state of $|0\rangle$ and $|1\rangle$ with the probability of α^2 and β^2.

Quantum gates are basically quantum circuits which are operating on qubits. Quantum logic gates are reversible in nature, and theoretically it is defined using unitary matrix. Among different quantum logic gates, the quantum rotational gate is used in this article which is represented as

$$\begin{bmatrix} \alpha_i' \\ \beta_i' \end{bmatrix} = \begin{bmatrix} \cos\theta_i & -\sin\theta_i \\ \sin\theta_i & \cos\theta_i \end{bmatrix} \begin{bmatrix} \alpha_i \\ \beta_i \end{bmatrix} \tag{3}$$

where (α_i,β_i) and (α_i',β_i') are the ith qubit before and after updating; and θ_i represents the rotation angle between α_i and β_i which can be determined in different ways based on the experiment.

3 Proposed Methodology

In this article, a quantum counterpart of classical modified genetic algorithm-based FCM is introduced for color MRI image segmentation. Accurate segmentation for MRI images is very much needed for better treatment. Modified genetic algorithm [14] already enhances the global search optimality of conventional GA by modifying the population initialization and crossover section, and this modified version of GA is incorporated to conventional FCM to overcome the early convergence to local minima. Though FCM is widely used for MRI image segmentation but at the very initial point, a proper initial class-level definition is essential for better clustering, otherwise it may hamper the total segmentation result. So to get over from this problem and also to make FCM faster, the class levels generated by the modified genetic algorithm (MfGA) are employed to FCM. This modified GA-based FCM obviously gives better result than conventional FCM and also GA-based FCM, but this total procedure takes a huge computational time. To present this algorithm in a more convenient and realistic way, some quantum computing phenomena like qubit, entanglement, orthogonality, quantum rotational gate are induced to this classical method. This quantum counterpart speeds up the classical method as well as enriches the efficacy; thus when the optimized class levels produced by this quantum induced modified genetic algorithm (QIMfGA) are fed to FCM as initial class levels, then the total segmentation results overwhelm the result of classical method. We have used this QIMfGA-based FCM for color MRI image segmentation. A color image contains much more information than grayscale image; so only using a proper segmentation algorithm all the essential information of the image can be taken care during segmentation. A color image has three planes, i.e., Red (R), Green (G), and Blue (B). So at first we need to decompose the color image into three separated planes, then for each plane this proposed method is applied. After completion of the process, output of each color component is combined to get the final output.

3.1 Preprocessing

As the MRI images are formed with the help of magnetic fields, radio wave, and field gradients, irregular biasing of pixel intensity values mostly occur which leads to the noise incorporation in the image. So after getting MRI image, first we need to remove all the noise using different noise removal algorithm as possible. Now this noise-free image is considered for the segmentation purpose for better recognition of affected parts.

3.2 Quantum-Inspired Modified GA (QIMfGA)-Based FCM

The steps of QIMfGA-based FCM algorithm for color MRI image segmentation are described in below:

1. After taking a noise-free color MRI image, the first job is to decompose this color image into three different color components of Red (R), Green (G), and Blue (B). Now the following steps are employed to each color component separately.
2. Population initialization takes place for each color component. In population initialization, a fixed size of chromosome pool is generated for further processing.

These chromosomes are made up with the cluster centroids. In MfGA, chromosomes are generated by a random selection of pixel intensity values within its minimum and maximum intensity values. Thus, to partition an image into N number of segments, $N + 1$ number of cluster centroids are initially chosen. Now using a weighted mean formula (Eq. 4), ultimate N number of class levels are produced; thus the population P is formed.

$$N_i = \frac{\sum_{j=R_i}^{R_{i+1}} f_j * I_j}{\sum_{j=R_i}^{R_{i+1}} f_j} \tag{4}$$

where R_i and R_{i+1} are the temporary class levels; f_j is the frequency of the jth pixel, and I_j shows the intensity value of the jth pixel.

3. Each centroid of each chromosome of population P is now randomly transformed to a real number within the value of 0 and 1 to generate population called P'.
4. Afterward, quantum orthogonality property is applied to population P' to construct population P''.
5. Now for quick convergence quantum rotational gate is adjoined based on the rotational angle to each chromosome of population P'' and create the ultimate population pool P^+.
6. Like GA, fitness value of each chromosome of population P^+ is calculated based on some fitness functions. Using Roulette-Wheel selection procedure, the fitter chromosomes are selected for further processing where crossover and mutation properties are applied.
7. The selected chromosomes are crossovered with each other based on the crossover probability. Unlike GA, here crossover probability is varied with the number of iterations. To retain good chromosomes in the population pool, the crossover probability is maintained in this manner that it will decrease as the iterations are increased. It can be described as

$$C_{prob} = C_{max} - \frac{C_{max} - C_{min}}{Iter_{max} - Iter_{current}} \tag{5}$$

where C_{prob} refers to the crossover probability, C_{max} and C_{min} indicate the maximum and minimum crossover probabilities; $Iter_{max}$ and $Iter_{current}$ refer to the maximum iteration number and the current iteration number, respectively.

8. Mutation is the process to maintain the genetic diversity from one generation to next generation. The crossovered chromosomes, i.e., the child solutions are altered here in one or more gene values based on mutation probability.
9. This diversified child solutions are now mixed with the parent solutions to create population P' of next generation. Steps 4–9 are repeated for a certain number of iterations to get the optimized output class levels.

10. The globally optimized class levels produced by QIMfGA algorithm are fed to FCM as the initial class levels to bring down the local minima convergence problem. After completion of FCM algorithm, the desired optimized output segmented image for each color component (R, G, B) is revealed.
11. At the very last stage, all the segmented output images of each color component are assembled to get the final color MRI segmented image.

4 Experimental Results

Segmentation results of two color MRI images using QIMfGA-based FCM algorithm are reported in this section. Segmentation results are measured with the use of two standard empirical measures $F(I)$ [20] and $Q(I)$ [21]. These two evaluation metrics are used to measure the dissimilarity between original and segmented images. As an MRI image contains white matter, gray matter, cerebrospinal fluids, and different non-invasive cells, we segment an MRI image into 6 to 10 segments. But only 6-segmented results for each image are presented in this article. Experimental results compared with quantum-inspired GA-based FCM and classical MfGA-based FCM are also reported here. At the starting point, this is essential to define the value of some constant terms. In our algorithm, a fixed population of size 50 is considered throughout the whole process. In case of QIMfGA maximum and minimum

Table 1 Class boundaries and evaluated segmentation quality measures, $F(I)$ by different algorithms for different classes of MRI image1

Algorithm	Sl no.	Class levels	Fitness value
MfGA-based FCM	1	(R = 1,13,68,128, 190,245), (G = 10,36,96,152,187,233), (B = 12,21,28,66,102,136)	7.68E+10
	2	**(R = 1,13,62,130,186,245), (G = 14,38,101,156,192,236), (B = 2,15,33,66, 102,136)**	**5.78E+10**
	3	(R = 1,14,68,125,193,244), (G = 11,36,101,156,187,233), (B = 2,14, 31,65,105,133)	5.98E+10
QIGA-based FCM	1	(R = 1,13,68,129,191,245), (G = 10,32,83,135,170,229), (B = 2,15,31,65,102,135)	6.79E+09
	2	(R = 1,12,66,130,191,245), (G = 14,32,87,135,166,229), (B = 2,15,31,65,102,137)	7.19E+09
	3	**(R = 1,14,68,129,191,244), (G = 11,35,83,135,174,232), (B = 2,15,30,67,102,133)**	**5.89E+09**
QIMfGA-based FCM	1	(R = 1,13,68,132,191,245), (G = 11,35, 83,135,170,229), (B = 2,15,31,65,99,133)	3.34E+09
	2	**(R = 1,13,68,129,191,245), (G = 12,36,87,135,170,229), (B = 2,15,31,65,102,131)**	**1.32E+09**
	3	(R = 1,13,68,129,191,245), (G = 13,28,83,132,170,231), (B = 2,15,31,65,101,136)	2.23E+09

Table 2 Class boundaries and evaluated segmentation quality measures, $Q(I)$ by different algorithms for different classes of MRI image1

Algorithm	Sl no.	Class levels	Fitness value
MfGA-based FCM	1	(R = 1,13,69,128, 191,245), (G = 10,36,96,152,187,233), (B = 2,20,28,66,102,136)	15816.23
	2	**(R = 1,13,68,128, 190,245), (G = 9,36,96,151,187,234), (B = 2,15,28,67,102,139)**	**11323.84**
	3	(R = 1,13,68,130, 182,245), (G = 10,36,96,152,187,233), (B = 2,16,30,66,99,136)	13258.45
QIGA-based FCM	1	**(R = 1,13,68, 128,190,245), (G = 10,36,96,152,187,234), (B = 2,18,29,71, 102,133)**	**7555.36**
	2	(R = 1,13,68,128,190,245), (G = 11,36,98,152,187,233), (B = 2,15,28,60,102,136)	8787.59
	3	(R = 1,13,68,128,190,245), (G = 10,32,96,152, 187,233), (B = 12,21,28,66,102,137)	8125.21
QIMfGA-based FCM	1	(R = 1,13,68,128,190,245), (G = 10,36,96,155,187,233), (B = 2,16,28,68,102,136)	5221.87
	2	(R = 1,13, 68,128,190,245), (G = 10,36,99,152,187,233), (B = 2,17,31,70, 102,136)	4541.41
	3	**(R = 1,13,68,128,189,245), (G = 10,36,96,152,184,233), (B = 2,17,28,65,102,136)**	**3982.48**

Table 3 Class boundaries and evaluated segmentation quality measures, $F(I)$ by different algorithms for different classes of MRI image2

Algorithm	Sl no.	Class levels	Fitness value
MfGA-based FCM	1	**(R = 6,52,102,158,200,238), (G = 4,50,93,135,171,207), (B = 5,49, 98,139,181,223)**	**4.13E+10**
	2	(R = 6,52,102,158,200,238), (G = 4,50,93,136,169,207), (B = 5,49,98,138,181,223)	4.55E+10
	3	(R = 6,52,105,158,200,238), (G = 4,50,93,136,179,207), (B = 5,49,95,138,181,223)	5.57E+10
QIGA-based FCM	1	(R = 6,52,102,158,200,238), (G = 4,50,93,136,171,207), (B = 5,44,98,139,189,223)	5.89E+09
	2	(R = 6,52,102,158,200,238), (G = 4,50,93,136,171,207), (B = 5,49,95,142,181,223)	6.54E+09
	3	**(R = 6,52,102,158,200,238), (G = 4,50,93,136,171,205), (B = 5,49,95,138,181,223)**	**5.14E+09**
QIMfGA-based FCM	1	**(R = 6,52,102,158,200,238), (G = 4,50,93,136,171,207), (B = 5,45,97,138,179,224)**	**2.32E+09**
	2	(R = 6,54,102,158,200,238), (G = 4,50,93,136,171,207), (B = 5,44,95,138,181,223)	2.41E+09
	3	(R = 6,52,102,158,201,241), (G = 4,50,93,136,174,207), (B = 5,45,95,138,181,223)	3.04E+09

Table 4 Class boundaries and evaluated segmentation quality measures, $Q(I)$ by different algorithms for different classes of MRI image2

Algorithm	Sl no.	Class levels	Fitness value
MfGA-based FCM	1	(R = 6,52,102,158, 200,238), (G = 4,50,93,136,171,207), (B = 5,49,95,138,181,223)	9473.51
	2	(R = 6,55, 105,158,201,238), (G = 4,50,93,131,170,207), (B = 5,49,97,138, 181,223)	10232.11
	3	**(R = 6,52, 102,158,200,238), (G = 4,49,93,136,171,207), (B = 5,45,95,139, 181,224)**	**8967.32**
QIGA-based FCM	1	**(R = 7,52,100,158,200,240), (G = 4,50,93,136,171,207), (B = 6,49,98,138,181,223)**	**5876.44**
	2	(R = 6,52,102,160,199,239), (G = 4,50,93,136,171,207), (B = 5,49,95,140,179,223)	6547.1
	3	(R = 6,52,101,158,200,238), (G = 4,50,91,136,171,207), (B = 5,49, 95,138,181,223)	7825.68
QIMfGA-based FCM	1	(R = 6,55,102,161,200,238), (G = 4,50,93,132,171,209), (B = 5,49,99,138,181,223)	3698.25
	2	**(R = 6,52,102,158,200,238), (G = 4,50,93,136,171,207), (B = 5,44, 96,142,181,223)**	**3398.47**
	3	(R = 6,52,100,158,200,241), (G = 4,50,93,136,171,207), (B = 5,49,95,138,187,223)	4251.87

Table 5 Different algorithm-based mean and standard deviation using different types of fitness functions and mean of time taken by different algorithms for MRI image1

Fit. fn.	Algorithm	Mean ± Std. div.	Mean time
$F(I)$	MfGA-based FCM	5.71E+10±1.31E+10	00:02:10
	QIGA-based FCM	6.83E+09±7.91E+08	00:01:53
	QIMfGA-based FCM	**3.29E+09 ±1.21E+09**	**00:01:37**
$Q(I)$	MfGA-based FCM	13166.49±2356.43	00:02:16
	QIGA-based FCM	8125.51±883.48	00:01:57
	QIMfGA-based FCM	**4722.44±929.75**	**00:01:31**

Table 6 Different algorithm-based mean and standard deviation using different types of fitness functions and mean of time taken by different algorithms for MRI image2

Fit. fn.	Algorithm	Mean ± Std. div.	Mean time
$F(I)$	MfGA-based FCM	4.07E+10±8.51E+09	00:02:21
	QIGA-based FCM	5.61E+09±1.02E+09	00:02:04
	QIMfGA-based FCM	**3.55E+09±8.94E+08**	**00:01:46**
$Q(I)$	MfGA-based FCM	9553.17±574.11	00:02:26
	QIGA-based FCM	7678.66±1462.24	00:02:03
	QIMfGA-based FCM	**4389.64±914.89**	**00:01:51**

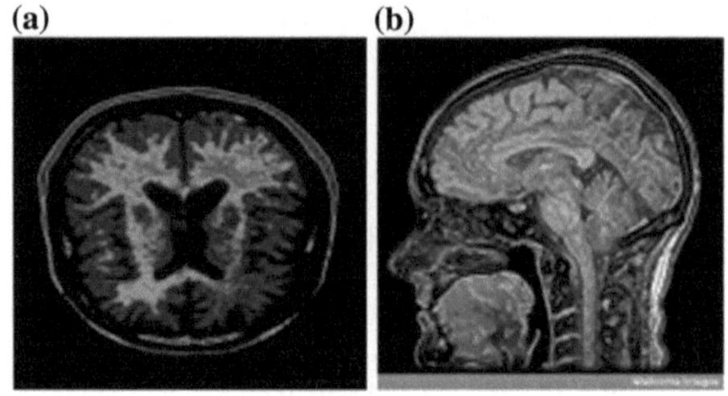

Fig. 1 **a** Color MRI image 1; **b** Color MRI image 2

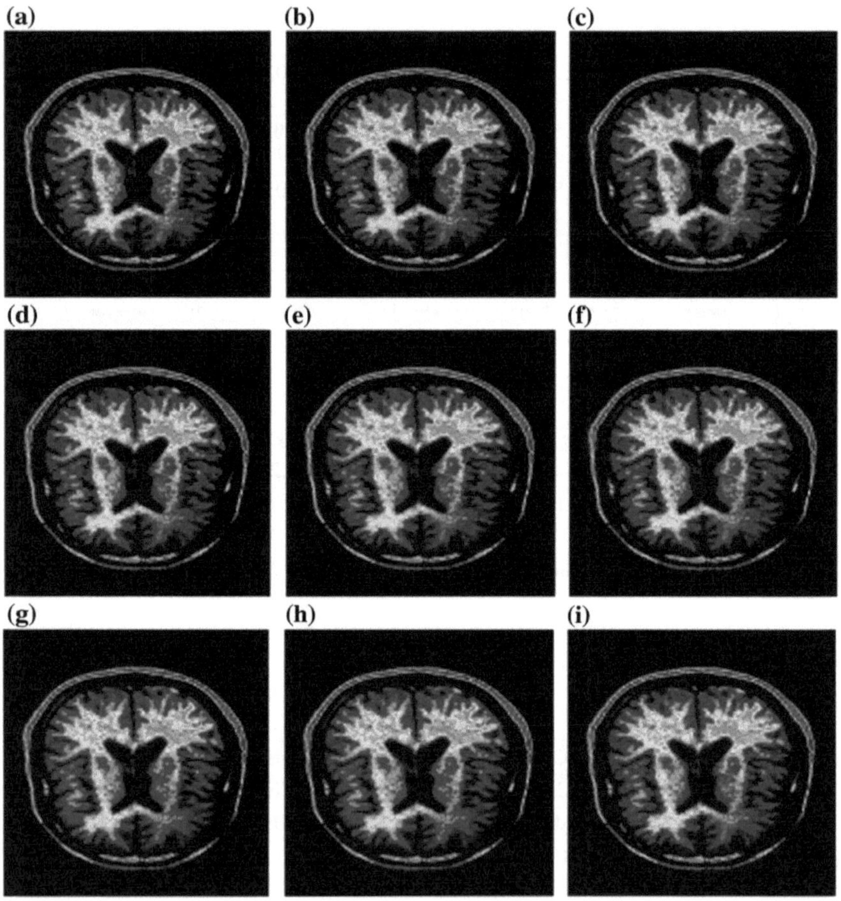

Fig. 2 6-class segmented 256 × 256 color MRI image1 with the class levels obtained by (**a**–**c**) MfGA-based FCM, (**d**–**f**) QIGA-based FCM, (**g**–**i**) QIMfGA-based FCM algorithm of three results of Table 1 with $F(I)$ as the quality measure

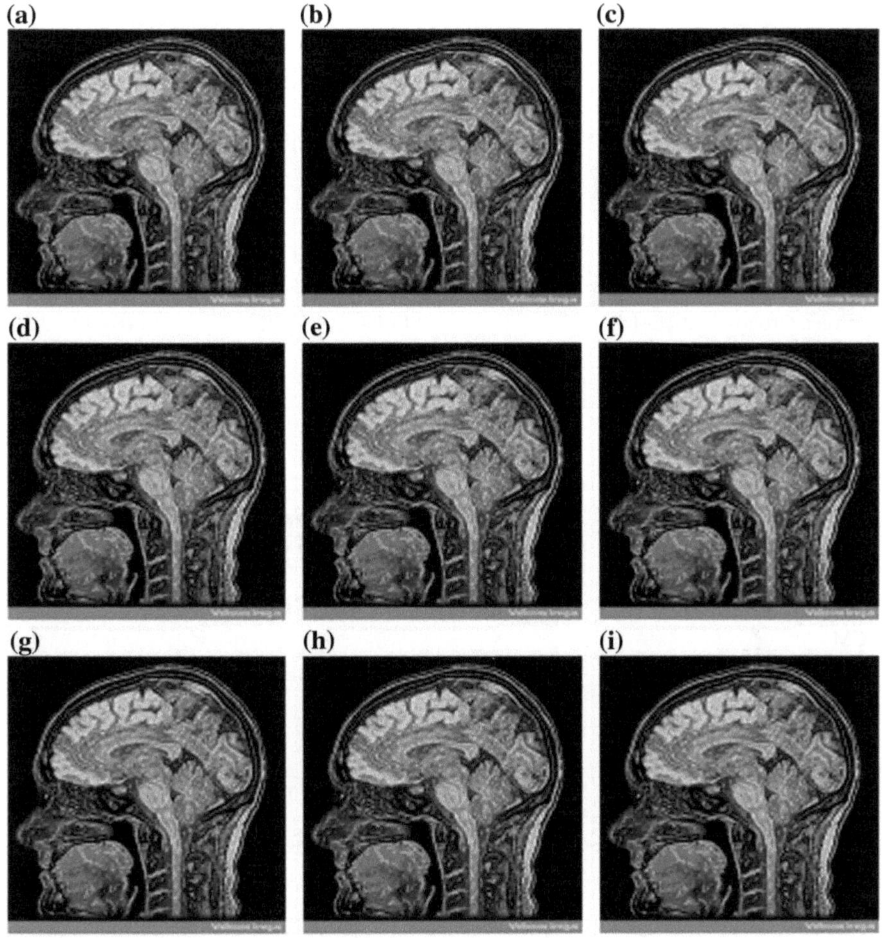

Fig. 3 6-class segmented 256 × 256 color MRI image2 with the class levels obtained by (**a**–**c**) MfGA-based FCM, (**d**–**f**) QIGA-based FCM, (**g**–**i**) QIMfGA-based FCM algorithm of three results of Table 4 with $Q(I)$ as the quality measure

crossover probability is applied as 0.9 and 0.5 respectively, based on which the present crossover probability for that instance of iteration is manipulated. Mutation takes place based on the mutation probability which is taken as 0.01. We have run all the methods on two test images several times, but here only three best results are presented for each methods. In Tables 1 and 2, segmentation results for MRI image1 are demonstrated for each methods based on $F(I)$ and $Q(I)$ respectively. The best results are boldfaced. In the same manner, in Tables 3 and 4 the class levels and value of measurement based on $F(I)$ and $Q(I)$ are illustrated. The mean, standard deviation and computational time taken for FCM part of each method of each image are reported in Tables 5 and 6,which ensures the efficacy of the proposed method.

Table 7 Single ANOVA analysis based on $Q(I)$ for MRI image1

Groups	Count	Sum	Average	Variance		
MfGA-based FCM	10	131664.93	13166.49	5.55E+06		
QIGA-based FCM	10	81255.15	8125.51	7.80E+05		
QIMfGA-based FCM	10	47224.45	4722.44	8.64E+05		
Source of Variation	**SS**	**df**	**MS**	**F**	**P-value**	F_{crit}
Between Groups	3.61E+08	2	1.8E+08	75.22	9.14E−12	3.35
Within Groups	6.47E+07	27	2399261			
Total	4.26E+08	29				

Table 8 Single ANOVA analysis based on $F(I)$ for MRI image2

Groups	Count	Sum	Average	Variance		
MfGA-based FCM	10	4.07E+11	4.07E+10	7.24E+19		
QIGA-based FCM	10	5.62E+10	5.62E+09	1.04E+18		
QIMfGA-based FCM	10	3.55E+10	3.55E+09	8E+17		
Source of Variation	**SS**	**df**	**MS**	**F**	**P-value**	F_{crit}
Between Groups	8.73E+21	2	4.36E+21	176.22	3.2E−16	3.35
Within Groups	6.69E+20	27	2.48E+19			
Total	9.4E+21	29				

Original and segmented images of both test images are presented in Figs. 1, 2, and 3 for each method. ANOVA analysis is also applied for the both images for each method. The anova results for each image are reported in Tables 7 and 8. All the presented facts and figures prove the efficiency of the proposed methodology.

5 Conclusion

A quantum-inspired modified genetic algorithm (QIMfGA)-based FCM clustering approach is discussed here for segmentation of color MRI images. This method has been compared with its classical counterpart and also with quantum-inspired genetic algorithm-based FCM. From all the accounted results, it is shown that the quantum version of MfGA-based FCM takes less computational time than its classical counterpart and also gives us a better result. It is also seen that this QIMfGA-based FCM performs well than QIGA-based FCM. All the reported results indicate superiority of the proposed methodology qualitatively and quantitatively for color MRI image segmentation.

References

1. Bezdek JC (1981) Pattern recognition with fuzzy objective function algorithms. Plenum, New York
2. Ahmed MN, Yamany SM, Mohamed N, Farag AA, Moriarty T (2002) A modified fuzzy C-means algorithm for bias field estimation and segmentation of MRI data. IEEE Trans Med Imaging 21(3):193–199
3. Yang Z, Chung FL, Shitong W (2009) Robust fuzzy clustering-based image segmentation. Appl Soft Comput 9(1):80–84
4. Adhikari SK, Sing JK, Basu DK, Nasipuri M (2015) Conditional spatial fuzzy C-means clustering algorithm for segmentation of MRI images. Appl Soft Comput 34:758–769
5. Chaabane SB, Bouchouicha M, Fnaiech F (2015) A hybrid technique for color image segmentation, application to the fire forest images. Int J Sci Eng Res (IJSER)
6. Tan KS, Isa NAM (2011) Color image segmentation using histogram thresholding Fuzzy C-means hybrid approach. Pattern Recognit 44:115
7. Chen M, Ludwig SA (2017) Color image segmentation using fuzzy C-regression model. Article ID, Adv Fuzzy Syst, p 4582948
8. De S, Bhattacharyya S, Chakraborty S (2010) True color image segmentation by an optimized multilevel activation function. In: IEEE international conference on computational intelligence and computing research. pp 545–548. https://doi.org/10.1007/978-1-4757-0450-1
9. De S, Bhattacharyya S, Chakraborty S. Color image segmentation using parallel OptiMUSIG activation function. Appl Soft Comput J 12:3228–3236
10. Amelio A, Pizzuti C (2013) A genetic algorithm for color image segmentation. In: Esparcia-Alcazar AI et al (eds) EvoApplications, LNCS 7835. Verlag, Berlin, Heidelberg, pp 314–323
11. Dong G, Xie M (2005) Color clustering and learning for image segmentation based on neural networks. IEEE Trans Neural Netw 16(4):925–936
12. Castillo O, Rubio E, Soria J, Naredo E (2012) Optimization of the fuzzy C-means algorithm using evolutionary methods. Eng Lett 20(1)
13. Mukhopadhyay A, Maulik U (2012) A multiobjective approach to MR brain image segmentation. Appl Soft Comput 11:872–880
14. Das S, De S (2016) Multilevel color image segmentation using modified genetic algorithm (MfGA) inspired fuzzy C-means clustering. In: 2nd International conference on research and computational intelligence and communication networks (ICRCICN), vol 1, pp 78–83
15. Talbi H, Draa A, Batouche M (2004) A new quantum-inspired genetic algorithm for solving the travelling salesman problem. In: 2004 IEEE international conference on industrial technology, KIT, vol 3, pp 1192–1197
16. Talbi H, Draa A (2017) A new real-coded quantum-inspired evolutionary algorithm for continuous optimization. Appl Soft Comput 61:765–791
17. Nebti S (2013) Bio-inspired algorithms for color image segmentation. Int J Comput Appl 73(18)
18. Caraiman S, Manta VI (2015) Image segmentation on a quantum computer. Quantum Inf Process 14(5):1693–1715
19. Dey S, Bhattacharyya S, Maulik U (2013) Quantum inspired meta-heuristic algorithms for multi-level thresholding for true colour images In: 2013 Annual IEEE conference on india conference (INDICON), pp 1–6

20. Liu J, Yang YH (1994) Multi-resolution color image segmentation. IEEE Trans Pattern Anal Mach Intel 16(7):689–700. https://doi.org/10.1109/34.297949
21. Borsotti M, Campadelli P, Schettini R (1998) Quantitative evaluation of color image segmentation results. Pattern Recognit Lett 19:741–747. https://doi.org/10.1016/S0167-8655(98)00052-X

Multi-verse Optimization Clustering Algorithm for Binarization of Handwritten Documents

Mohamed Abd Elfattah, Aboul Ella Hassanien, Sherihan Abuelenin and Siddhartha Bhattacharyya

Abstract Binarization process of images of historical manuscripts is considered a challenge due to the different types of noise that are related to the degraded manuscripts. This paper presents an automatic clustering algorithm for binarization of handwritten documents (HD) based on multi-verse optimization. The multi-verse algorithm is used to find cluster centers in HD where the number of clusters is predefined. The proposed approach is tested on the benchmarking dataset used in the Handwritten Document Image Binarization Contest (H-DIBCO 2014). The proposed approach is assessed through several performance measures. The experimental results achieved competitive outcomes compared to the well-known binarization methods such as Otsu and Sauvola.

Keywords Binarization · Handwritten documents · Multi-verse optimizer (MVO) H-DIBCO 2014

1 Introduction

Binarization in document analysis field is considered as an open challenge, since the historical manuscript images suffer from different kinds of noise and any proposed

M. A. Elfattah · S. Abuelenin
Faculty of Computers and Information, Computer Science Department, Mansoura University, Dakahlia Governorate, Egypt

A. E. Hassanien
Faculty of Computers and Information, Cairo University, Giza, Egypt

S. Bhattacharyya (✉)
Department of Computer Application, RCC Institute of Information Technology, Kolkata 700015, India
e-mail: dr.siddhartha.bhattacharyya@gmail.com
URL: http://www.egyptscience.net

M. A. Elfattah · A. E. Hassanien · S. Bhattacharyya
Scientific Research Group in Egypt (SRGE), Cairo, Egypt

S. Bhattacharyya et al. (eds.), *Recent Trends in Signal and Image Processing*, Advances in Intelligent Systems and Computing 727,
https://doi.org/10.1007/978-981-10-8863-6_17

systems, such as optical character recognition (OCR) and word spotting (WS), need the proper binarized image, while the accuracy of these systems affected directly with this process (binarization). Binarization is the process of extracting the text without any noise (black) and background (white) [1]. With more kinds of noise as; smudge, multi-colored, bleed-through, unclear background, shadow, broken character. The current systems depend on the binarization as the first step which is affected by noise. These systems are included in many applications such as handwritten recognition, watermarking, and data hiding [2].

Thresholding approaches can be either local or global. In the case of degraded images, global approaches do not perform well [3]. Otsu [4], Kapur et al. [5], and Kittler and Illingworth [6] are considered global methods, while Niblack [7], Sauvola and Pietikäinen [8], and Bernsen [9] are considered local methods. From the literature, many different approaches are presented to binarize the degraded images. However, the binarization process is still an open challenge [10].

Recently, meta-heuristic optimization algorithms have a wide range of applications such as feature selection, image processing, and others. Nature-inspired algorithms are well-known optimization algorithms. In these algorithms, the local optima problem can be solved by sharing the information between candidates [11]. Therefore, in this paper, a new cluster algorithm is proposed using one of the recent optimization algorithms named multi-verse optimizer (MVO) [11]. This algorithm is proposed to address the binarization process of historical documents.

The rest of this paper is organized as follows: Section 2 introduces the basics of MVO algorithm. Section 3 presents the proposed approach. In Sect. 4, the experimental result and discussion are clarified. Finally, conclusions and future works are presented in Sect. 5.

2 Preliminaries: Multi-verse Optimizer (MVO)

MVO is a recent nature-inspired algorithm proposed by Mirjalili et al. [11]. It is based on the three concepts of cosmology (white hole, black hole, and wormhole). The exploration phase is based on (white, black hole), while the wormhole is employed for improving the quality in the exploitation phase [11].

At each iteration, these universes are sorted depending on their inflation rate. The roulette wheel is employed for the selecting to have a white hole:

$$\mathbf{U} = \begin{bmatrix} y_1^1 & \cdots & y_1^v \\ \cdots & \cdots & \cdots \\ y_n^1 & \cdots & y_n^v \end{bmatrix} \tag{1}$$

where the number of parameters (variables) is presented by v and the number of universes by n.

$$y_i^j = \begin{cases} y_k^j & r1 < NI(Ui) \\ y_i^j & r1 \geq NI(Ui) \end{cases} \tag{2}$$

where y_i^j denotes jth of ith universes. Ui presents the ith universe. The normalized inflation rate is presented by $NI(Ui)$ of the ith universe, $r1$ is a random value in [0, 1], and y_k^j presents the jth parameter of kth universe chosen by a roulette wheel selection mechanism [11].

To update the solutions, the two parameters Traveling Distance Rate (TDR) and Wormhole Existence Probability (WEP) are calculated based on Eqs. 3 and 4:

$$WEP = min + l \times \left(\frac{max - min}{L}\right) \tag{3}$$

The minimum and maximum are presented by min (0.2) and max (1) as in Table 1, respectively, while the current iteration presented by l and L denotes the maximum number of iterations:

$$TDR = 1 - \left(\frac{l^{1/p}}{L^{1/p}}\right) \tag{4}$$

The exploitation accuracy is presented by p. The large value of p indicates high perfect of local search/exploitation. The position of solutions is updated based on Eq. 5:

$$y_i^j = \begin{cases} \begin{cases} Y_j + \text{TDR} \times \left(\left(ub_j - lb_j\right) \times r4 + lb_j\right) & r3 < 0.5 \\ Y_j - \text{TDR} \times \left(\left(ub_j - lb_j\right) \times r4 + lb_j\right) & r3 \geq 0.5 \end{cases} & r2 < \text{WEP} \\ y_i^j & r2 \geq \text{WEP} \end{cases} \tag{5}$$

where Yj denotes the jth parameter of the best universe; lb_j and ub_j denote the lower and upper bound of jth variable, while, r2, r3, and r4 are random numbers in [0, 1]. y_j^i denotes the jth parameter of ith universe. TDR and WEP are coefficients [11].

3 The Proposed Binarization Approach

Starting with applying the MVO algorithm on the degraded manuscripts image to find the optimal cluster center based on objective function given in Eq. 6 as in the basic K-means clustering algorithm [12]. Depending on the obtained cluster centers

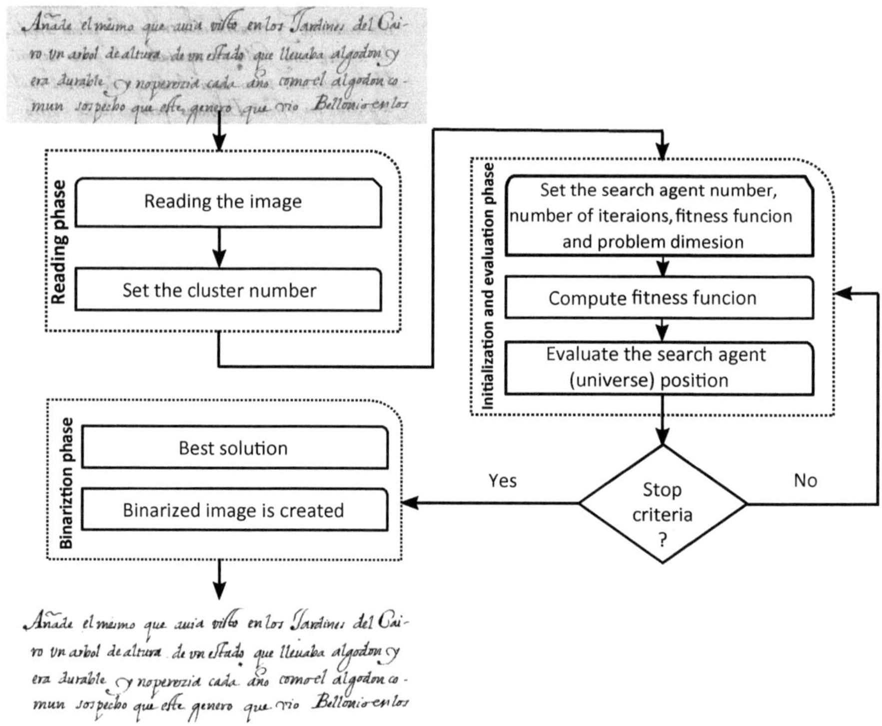

Fig. 1 General architecture of the proposed binarization approach

to create BW (white, black) representing the foreground by white pixels where the darkest cluster denotes the text. In fact, at every iteration, each (universe) search agent updates its position according to (the best position). Finally, the cluster centers are updated, and the binary image is created. Figure 1 illustrates the general architecture of the proposed binarization approach and its phases.

3.1 *Fitness Function and MVO Parameters*

Equation 6 is the squared error function that is used as an objective function of the multi-verse optimization algorithm typically as in the *k*-means clustering [12]:

Table 1 MVO parameter's setting

Parameter	Value
Number of universes	8
Max iterations	10
Min (Eq. 3)	0.2
Max (Eq. 3)	1
No. of run	30
No. of clusters	2
Dimension	1
Range	[0 255]

$$J = \sum_{j=1}^{k} \sum_{i=1}^{x} \left\| x_i{}^{(j)} - c_j \right\|^2 \quad (6)$$

The distance measure among the cluster center c_j and data points $x_i{}^{(j)}$ is presented by $\left\| x_i{}^{(j)} - c_j \right\|^2$. It denotes the distance of the n data points from their cluster centers.

The foremost target of MVO is to minimize this function. Each cluster is presented within a single centroid. Each universe presents one solution, and its position is updated according to (best solution). For any optimization algorithm, we primarily require setting some parameters value that provides better performance of the proposed approach. Table 1 refers to the MVO parameters setting.

4 Experimental Results and Discussion

H-DIBCO 2014 dataset [13] is used and employed to evaluate the proposed approach. This dataset contains ten handwritten images with different kinds of noise which are collected from *tranScriptorium* project [14]. This dataset is available with its ground truth. This dataset contains illustrative degradations such as bleed-through, faint characters, smudge, and low contrast.

To evaluate the proposed approach, different performance measures [15] are used and employed including F-measure [16, 17], Negative Rate Metric (NRM) [18], Peak Signal-to-Noise Ratio (PSNR) [17], Distance Reciprocal Distortion (DRD) [2], and Misclassification Penalty Metric (MPM) [18, 19]. The high values of F-measure, PSNR, and low value on DRD, NRM, and MPM indicate the best result. In addition, visual inspection is used.

Table 2 Result of MVO on H-DIBCO 2014

Image name	F-measure	PSNR	DRD	NRM	MPM
H01	91.79	20.47	2.05	0.05	0.08
H02	89.48	17.56	2.93	0.06	0.45
H03	98.09	24.04	0.92	0.01	0.03
H04	94.94	18.32	1.60	0.04	0.37
H05	94.38	17.47	2.04	0.04	0.66
H06	94.05	17.56	2.36	0.04	0.51
H07	84.85	15.32	5.98	0.04	7.97
H08	94.12	24.59	1.51	0.04	0.14
H09	88.77	16.99	2.72	0.09	0.21
H10	89.66	17.30	2.51	0.08	0.28
Average	92.01	18.96	2.46	0.04	1.07

Table 3 Results of the MVO with the state-of-the-art methods on H-DIBCO 2014

Approach name	F-measure	PSNR	DRD
1 [13]	96.88	22.66	0.902
2 [13]	96.63	22.40	1.001
3 [13]	93.35	19.45	2.194
4 [13]	89.24	18.49	4.502
Otsu [13]	91.78	18.72	2.647
Sauvola [13]	86.83	17.63	4.896
MVO result	92.01	18.96	2.46

Table 2 presents the results of MVO on H-DIBCO 2014; the high PSNR value appears in H08 with value 24.59, while the worst value is in H07 with value 15.32. The higher value of F-measure (98.09) is in H03, while the worst is in H07 (84.85). In addition, the better DRD value is in H03 (0.92). The best NRM and MPM appear in H03 (0.01, 0.03), respectively.

Table 3 summarizes the comparison between the approaches submitted to H-DIBCO 2014 competition [13] and the proposed MVO algorithm. According to Table 3, the numbers (1 to 4) indicate the rank of submitted methods with their values of F-measure, PSNR, and DRD. The result of MVO is better than the well-known methods (Otsu and Sauvola) and the method number (4) in all performance measures.

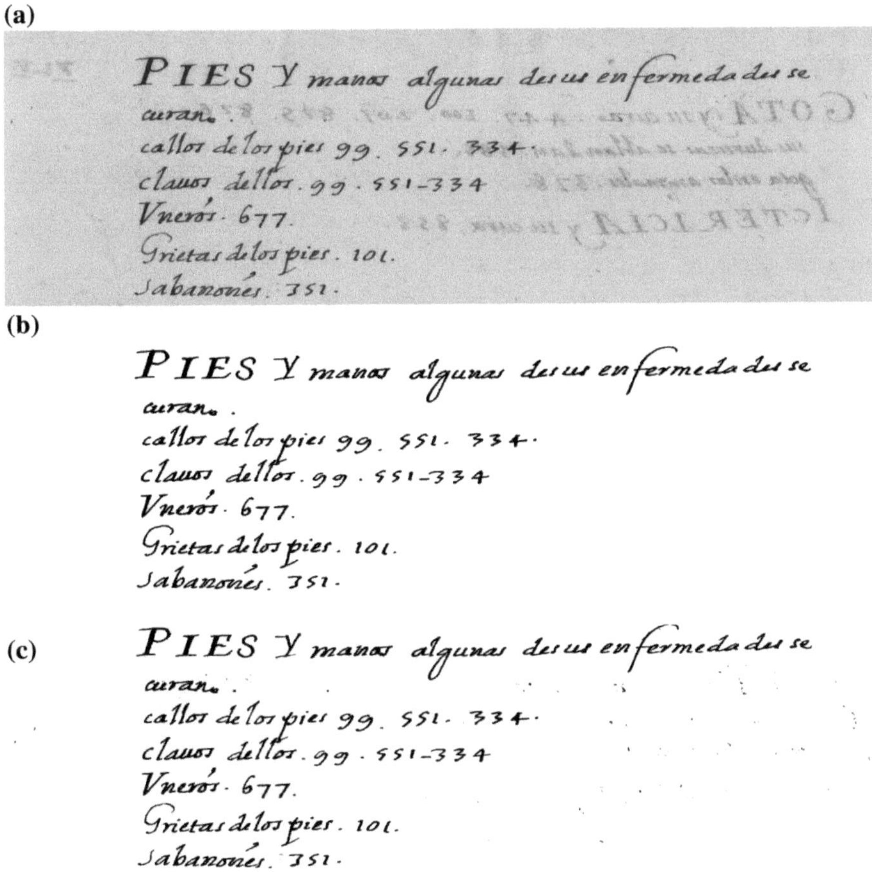

Fig. 2 **a** H08 (H-DIBCO 2014) test sample, **b** GT, and **c** MVO result

For visual inspection, two images are selected named H08 and H10 as shown in Figs. 2 and 3. Figures 2 and 3 show the comparison between the ground truth images, as shown in Figs. 2b and 3b, and the MVO output images (Figs. 2c and 3c). From these figures, the output images are very close to the ground truth images with complete character structure, but we found some simple noise.

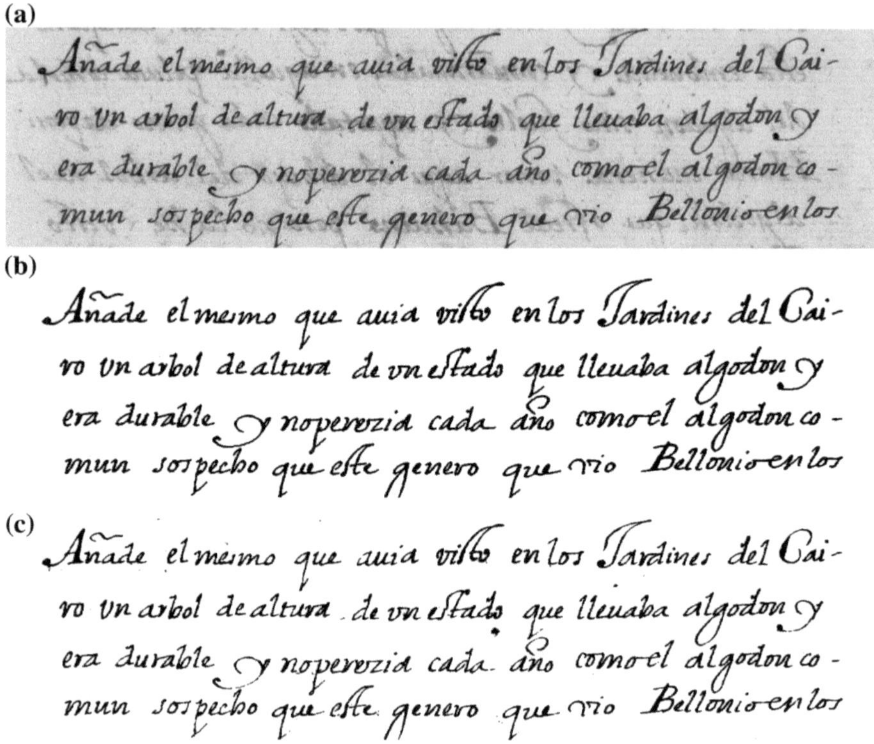

Fig. 3 **a** H10 (H-DIBCO 2014) test sample, **b** GT, and **c** MVO result

The convergence rate is the last judgment measure to evaluate the proposed binarization approach. In each iteration, the solution with the best fitness is kept and it is used to create the convergence curves as in Fig. 4. This figure presents the convergence curve for two different images, and the lower fitness value with increasing the number of iterations demonstrates the convergence of the proposed approach. It is also remarkable that the fitness value decreased dramatically. The optimization problem here is a minimization problem. We can conclude from this figure that the MVO is a promising approach to address the binarization problem.

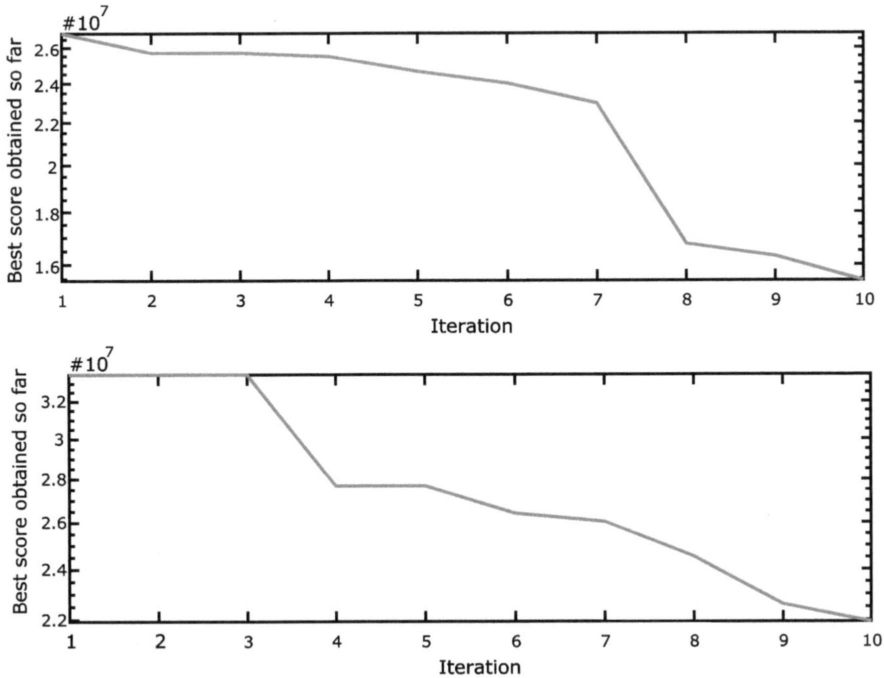

Fig. 4 MVO convergence curve

5 Conclusions and Future Works

This paper presents a binarization approach based on the MVO algorithm, which is employed for minimizing the distance between clusters. The convergence curve rate proves the high speed of MVO algorithm. This approach can deal with various kinds of noise.

As future work, it is planned to use preprocessing phase which can improve the accuracy of binarization. Furthermore, hybridization with other optimization algorithms will be used to improve the results in [20–24]. A comparative analysis between the basic MVO and a chaotic version of it based on different chaos maps and different objective functions will be presented to improve the OCR recognition rate in [25, 26] by using it in the binarization phase.

References

1. Mesquita RG, Silva RM, Mello CA, Miranda PB (2015) Parameter tuning for document image binarization using a racing algorithm. Expert Syst Appl 42(5):2593–2603
2. Lu H, Kot AC, Shi YQ (2004) Distance-reciprocal distortion measure for binary document images. IEEE Signal Process Lett 11(2):228–231
3. Singh BM, Sharma R, Ghosh D, Mittal A (2014) Adaptive binarization of severely degraded and non-uniformly illuminated documents. Int J Doc Anal Recognit (IJDAR) 17(4):393–412
4. Otsu N (1979) A threshold selection method from gray-level histograms. IEEE Trans Syst Man Cybern 9(1):62–66
5. Kapur JN, Sahoo PK, Wong AK (1985) A new method for gray-level picture thresholding using the entropy of the histogram. Comput Vis Graph Image Process 29(3):273–285
6. Kittler J, Illingworth J (1986) Minimum error thresholding. Pattern Recognit 19(1):41–47
7. Niblack W (1985) An introduction to digital image processing. Strandberg Publishing Company
8. Sauvola J, Pietikäinen M (2000) Adaptive document image binarization. Pattern Recognit 33(2):225–236
9. Bernsen J (1986) Dynamic thresholding of grey-level images. Int Conf Pattern Recognit 2:1251–1255
10. Hadjadj Z, Cheriet M, Meziane A, Cherfa Y (2017) A new efficient binarization method: application to degraded historical document images. Signal Image Video Process 1–8
11. Mirjalili S, Mirjalili S, Hatamlou A (2016) Multi-verse optimizer: a nature-inspired algorithm for global optimization. Neural Comput Appl 27(2)
12. MacQueen J et al (1967) Some methods for classification and analysis of multivariate observations. In: Proceedings of the fifth Berkeley symposium on mathematical statistics and probability, Oakland, CA, USA, vol 1, pp 281–297
13. Ntirogiannis K, Gatos B, Pratikakis I (2014) ICFHR2014 competition on handwritten document image binarization (h-dibco 2014). In: 2014 14th international conference on frontiers in handwriting recognition (ICFHR). IEEE, pp 809–813
14. http://transcriptorium.eu
15. Gatos B, Ntirogiannis K, Pratikakis I (2009) ICDAR 2009 document image binarization contest (DIBCO 2009). In: 10th international conference on document analysis and recognition, 2009 (ICDAR'09). IEEE, pp 1375–1382
16. Sokolova M, Lapalme G (2009) A systematic analysis of performance measures for classification tasks. Inf Process Manag 45(4):427–437
17. Ntirogiannis K, Gatos B, Pratikakis I (2013) Performance evaluation methodology for historical document image binarization. IEEE Trans Image Process 22(2):595–609
18. Pratikakis I, Gatos B, Ntirogiannis K (2010) H-dibco 2010-handwritten document image binarization competition. In: 2010 international conference on frontiers in handwriting recognition (ICFHR). IEEE, pp 727–732
19. Young DP, Ferryman JM (2005) Pets metrics: on-line performance evaluation service. In: Joint IEEE international workshop on visual surveillance and performance evaluation of tracking and surveillance (VS-PETS), pp 317–324
20. Elfattah MA, Abuelenin S, Hassanien AE, Pan JS (2016) Handwritten arabic manuscript image binarization using sine cosine optimization algorithm. In: International conference on genetic and evolutionary computing. Springer, pp 273–280
21. Mostafa A, Fouad A, Elfattah MA, Hassanien AE, Hefny H, Zhu SY, Schaefer G (2015) Ct liver segmentation using artificial bee colony optimisation. Procedia Comput Sci 60:1622–1630
22. Mostafa A, Elfattah MA, Fouad A, Hassanien AE, Hefny H (2016) Wolf local thresholding approach for liver image segmentation in ct images. In: Proceedings of the second international Afro-European conference for industrial advancement (AECIA 2015). Springer, pp 641–651
23. Ali AF, Mostafa A, Sayed GI, Elfattah MA, Hassanien AE (2016) Nature inspired optimization algorithms for ct liver segmentation. In: Medical imaging in clinical applications. Springer, pp 431–460

24. Hassanien AE, Elfattah MA, Aboulenin S, Schaefer G, Zhu SY, Korovin I (2016) Historic handwritten manuscript binarisation using whale optimisation. In: 2016 IEEE international conference on systems, man, and cybernetics (SMC). IEEE, pp 003842–003846
25. Sahlol AT, Suen CY, Zawbaa HM, Hassanien AE, Elfattah MA (2016) Bio-inspired bat optimization algorithm for handwritten arabic characters recognition. In: 2016 IEEE congress on evolutionary computation (CEC). IEEE, pp 1749–1756
26. Sahlol A, Elfattah MA, Suen CY, Hassanien AE (2016) Particle swarm optimization with random forests for handwritten arabic recognition system. In: International conference on advanced intelligent systems and informatics. Springer, pp 437–446

3D Object Recognition Based on Data Fusion at Feature Level via Principal Component Analysis

Abdelhameed Ibrahim, Aboul Ella Hassanien and Siddhartha Bhattacharyya

Abstract In computer vision applications, recognizing a 3D object is considered an effective research area. The recognition of 3D objects can be accomplished by building accurate 3D models of these objects. The 3D object reconstruction is the creation of 3D models from a set of images. Different reconstruction techniques can be used to obtain 3D models using calibrated and uncalibrated images of the required object. In this work, a method for evaluating the effectiveness of using 3D models produced from different 3D modeling techniques is presented. Different feature extraction, matching, and fusion techniques of average, addition, principal component analysis, and discrete wavelet transform are used to get multiple models. The targeted object models are then used as input to an artificial neural network which is used to compute the recognition error value. Experiments use real-world images for modeling to evaluate the proposed method.

Keywords Object recognition · Principal component analysis · Data fusion

1 Introduction

The use of three-dimensional (3D) data becomes increasingly popular in computer vision applications. The 3D object recognition is defined as the problem of deter-

---

A. Ibrahim
Faculty of Engineering, Mansoura University, Mansoura, Egypt

A. Ella Hassanien
Faculty of Computers and Information, Cairo University, Cairo, Egypt

S. Bhattacharyya (✉)
Department of Computer Application, RCC Institute of Information Technology, Kolkata 700 015, India
e-mail: dr.siddhartha.bhattacharyya@gmail.com
URL: http://www.egyptscience.net

A. Ibrahim · A. Ella Hassanien · S. Bhattacharyya
Scientific Research Group in Egypt, (SRGE), Giza, Egypt

S. Bhattacharyya et al. (eds.), *Recent Trends in Signal and Image Processing*, Advances in Intelligent Systems and Computing 727,
https://doi.org/10.1007/978-981-10-8863-6_18

mining the similarity between the model stored in a database and the scene object. The object recognition system main objectives are to detect and to recognize objects in an image as humans can do. We can recognize objects in images even changes in illumination or pose happen; also, we can analyze images taken from the left and right eyes to build a 3D model of an object. However, the computer system performs object modeling and recognition with more challenges than humans [1].

One of the most efficient techniques for 3D object recognition system is to reconstruct accurate and complete 3D models of these objects. This can be achieved by capturing some images of the targeted object from different viewpoints and then applying the processes of preprocessing, feature extraction and matching, and 3D model reconstruction. Preprocessing operations such as filtering [2], histogram equalization [3], or high dynamic range imaging [4] are performed. Features can be then extracted from images and match the corresponding features to build a 3D model [5, 6]. The scale-invariant feature transform (SIFT) [7, 8], segmentation-based fractal texture analysis (SFTA) [9], block matching (BM) [10], and corner detector (CD) [11] algorithms can be used for feature extraction.

There are two main techniques for 3D reconstruction: calibrated reconstruction [12], in which intrinsic and extrinsic parameters of the camera are obtained, and uncalibrated reconstruction [13], in which the camera calibration is not needed. The calibration parameters can help to get rectified images which can be used to compute a disparity map from which the 3D model can be reconstructed. However, in the uncalibrated reconstruction, all objects are maintained at a specified distance from the viewer. This distance can be computed by calculating the horizontal offset between matched points of the image pair. Depth information can be obtained by multiplying the horizontal offset by a gain. Depending on depth information, a 3D model can be reconstructed.

Researchers applied different techniques for recognizing 3D objects. Liang-Chia et al. [14] introduced a recognition method by computing the distance between feature vectors. Jie et al. [15] presented an object recognition method in which additional sensing and calibration information are leveraged together with large amounts of training data. Authors in [16] present a method to address, in the context of RGB-D (depth) cameras, joint object category, and instant recognition. An approach for object recognition was presented by Eunyoung and Gerard [17] that depends on maximizing the use of visibility context to boost dissimilarity between similar-shaped background objects and queried objects in the scene. Liefeng et al. [18] presented a kernel principal component analysis (KPCA) in which three types of match kernels are used to measure similarities between image patches. Kuk-Jin et al. [19] proposed a framework for general 3D object recognition. This was depending on using invariant local features and their 3D information with stereo cameras. Another method for solving and formulating the recognition problem for polyhedral objects was discussed in [20]. First, multiple cameras have been used to take multiple views from different directions. Then, a set of features were extracted from images and were used to train an artificial neural network (ANN) which was used for matching and classification of the 3D objects.

In this paper, a method for evaluating the effectiveness of using 3D models from different 3D modeling techniques in the recognition process is presented. Multiple feature extraction, matching features, and data fusion techniques are used to get the 3D models. For feature extraction, SIFT, BM, and CD algorithms are tested. Fusion techniques of average, addition, PCA, and discrete wavelet transform (DWT) are used. Fusion at feature level is applied. 3D models of the targeted object are then used as input to ANN which is used to compute the value of recognition error. Experiments use real-world images for modeling to evaluate the proposed method.

2 3D Model Reconstruction Techniques

Two main 3D reconstruction techniques of calibrated reconstruction and uncalibrated reconstruction exist. Camera calibration is performed to obtain parameters of the camera in calibrated techniques, while camera calibration is not needed in the uncalibrated one [18, 21, 22]. In the calibrated based approach, two images representing left and right views are taken by two calibrated cameras. Basic block matching is performed by extracting block around every pixel in one image and search in another image for block matching. The depth map is constructed with positive and negative disparities, and the map will be saturated to have only positive values. For each pixel, the optimal disparity is selected based on its cost function. Finally, the noise will be removed, and the foreground objects are being better reconstructed by performing dynamic programming. The 3D world coordinates are then computed.

The common uncalibrated reconstruction methods [13] are used for 3D model reconstruction. A camera is used to take object's images from more than a viewpoint. A regular handheld camera can be used for this purpose. Using different feature extraction techniques to extract features from images, then, match these features to find the corresponding 3D model. Several algorithms can be used for feature extraction and matching [8, 10, 11]. Depth information can be obtained according to the fact that a camera moves in the x-direction. The first step in calculating depth information is to compute the offset "H" of matched points. The horizontal offset is defined as the horizontal distance moved by an object between the left and right input images [13]. Depth can be obtained by multiplying the horizontal offset by a gain "A" as $D = A \times H$. Repeat this process for all the matched points; depth information can be obtained for the whole image to create a 3D surface of the object.

3 Proposed 3D Object Recognition Method

The purpose of object recognition using the 3D models is to recognize objects that are found in images. The proposed framework produces multiple 3D models to be used as targets for a feed-forward backpropagation ANN which have a hidden layer, an output layer, and ten neurons. Different images that contain objects with different positions will be used as inputs to the ANN as illustrated in the proposed approach general architecture shown in Fig. 1.

3.1 Feature Extraction and Matching Algorithms

One of the most important stages for reconstructing the 3D model is feature extraction and matching stage. There are different methods for extracting features and matching these features. Three methods of SIFT, block matching, and corner detector are considered in this work. The SIFT method has the steps of localizing keypoint and descriptor, assignment orientation, and detection of scale space extreme [7]. Block matching method is used to match two images by comparing blocks from both images. The two images are first represented as matrices of sub-blocks and then compare each sub-block with its adjacent sub-block. The sub-block is usually a square of side pixels [10]. Matching is achieved between sub-blocks when the output of the

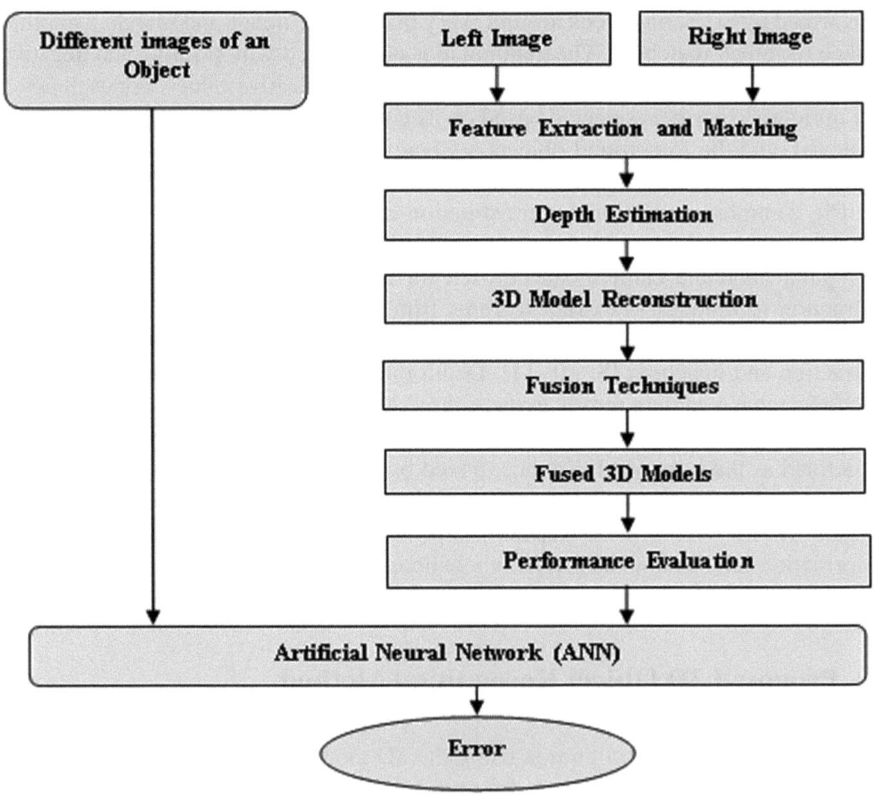

Fig. 1 Proposed 3D object recognition approach general architecture

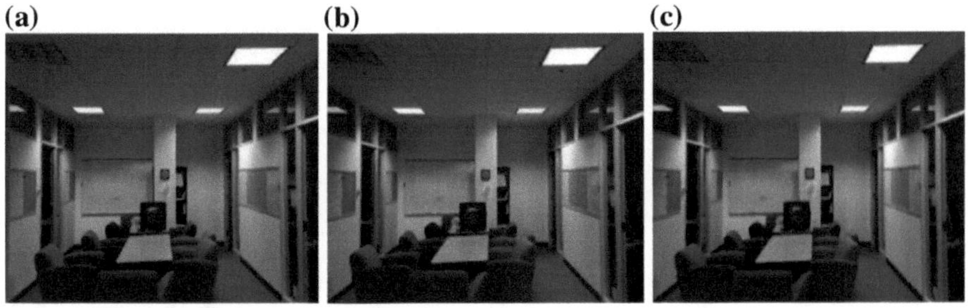

Fig. 2 Color images; **a** image1, **b** image2, **c** reference 3D model

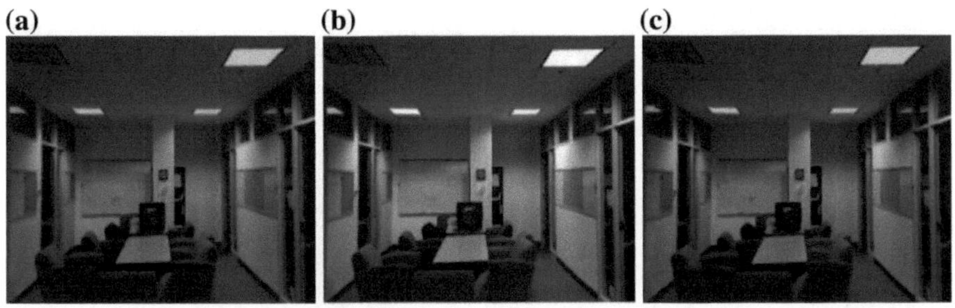

Fig. 3 Calibrated 3D models; **a** BM-based model, **b** SIFT-based model, and **c** CD-based model

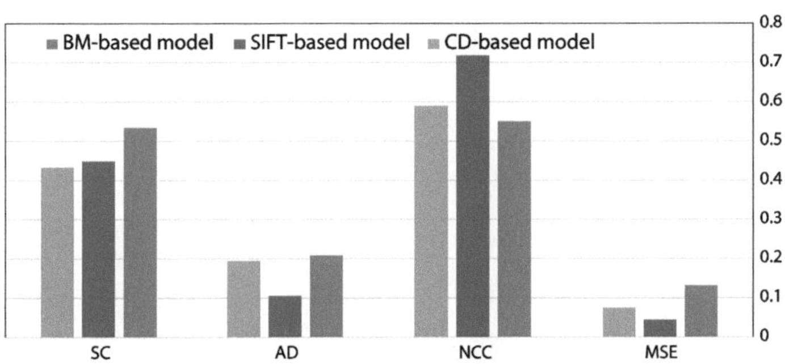

Fig. 4 Comparison between calibrated 3D models based on BM, SIFT, and CD

cost function, such as mean absolute difference (MAD) function, gives the least cost. Corner detector is a popular method which can be used for image matching, localization, tracking, and recognition purposes. The intersection of two edges is defined as a corner [11]. The Harris detector technique is used to extract corner point locations. Determining the values of the sum of squared differences in all directions is the goal of the corner detector technique. This method is based on calculating eigenvalues, and according to the calculated eigenvalues, it can be indicated that there is an edge or a corner.

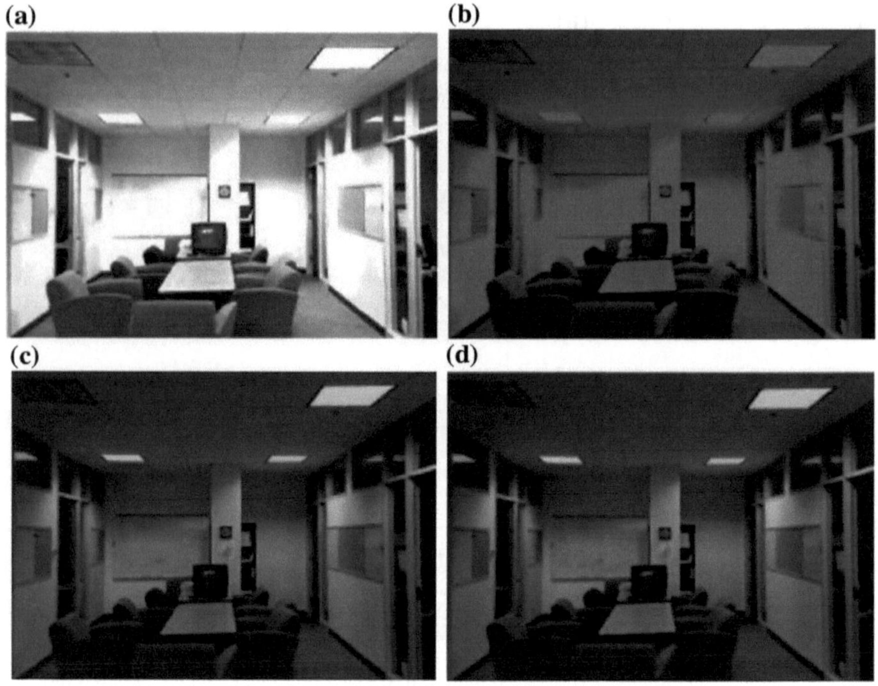

Fig. 5 Calibrated 3D models based on data fusion between SIFT-based model and CD-based model via: **a** addition, **b** average, **c** PCA, and **d** DWT

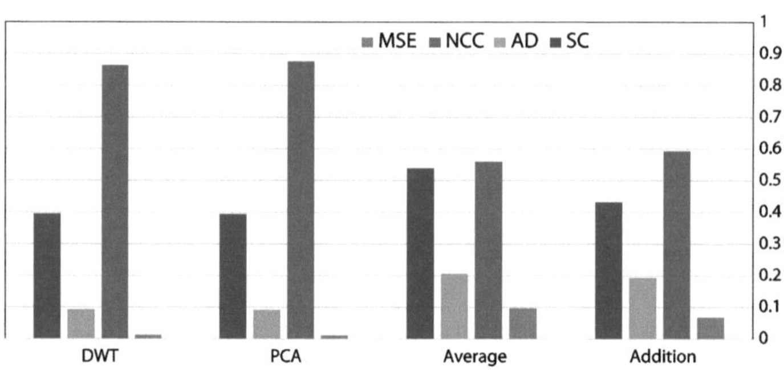

Fig. 6 Results of 3D models by data fusion of SIFT-based model and CD-based model

3.2 *Fusion Techniques*

Fusion techniques are used to produce the final 3D models. The common methods tested here are addition, average, PCA [23–25], and DWT [26]. The fusion using addition/average techniques is simply computed the fused 3D model pixel value by adding/averaging values of pixels in both 3D models. In PCA, a mathematical model is used to transform correlated variables into uncorrelated variables. Finally,

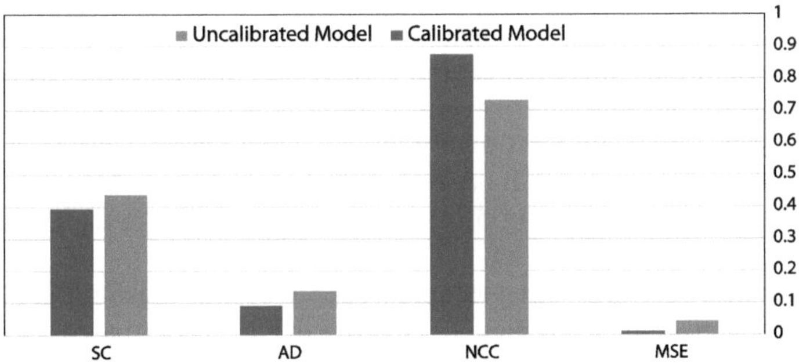

Fig. 7 Comparison between PCA-based 3D models resulted from applying calibrated and uncalibrated reconstruction techniques

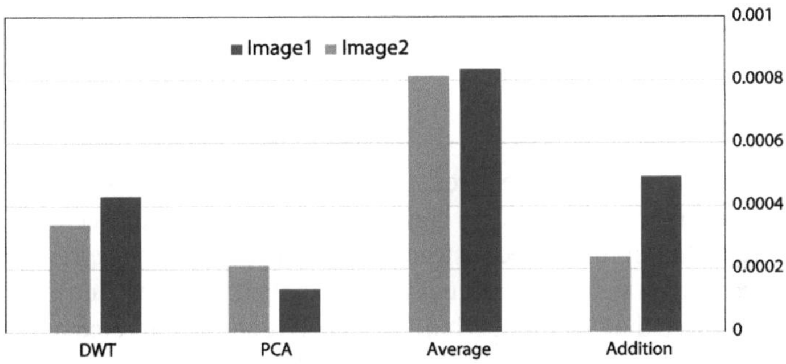

Fig. 8 Recognition error results for calibrated reconstruction using different fusion techniques

DWT is used to compute coefficients from a wavelet series expansion which maps function of a continuous variable into coefficients' sequence. The fused image is then reconstructed by the inverse DWT (IDWT) [13].

4 Experimental Results

Two test color images with a resolution of $300 \times 400 \times 3$ are used for the proposed method evaluation. The two images are shown in Fig. 2 with reference image shown in Fig. 2c. Performance metrics are usually used between the reconstructed 3D model from different methods and the reference 3D model in the database which can then be used for the object recognition process. In this work, the quality of the resulted 3D models is measured using the metrics of average difference (AD), structural content (SC), normalized cross-correlation (NCC), and mean square error (MSE) [27]. The accuracy of the recognition process is inversely proportional to the value of error produced by the neural network [28].

Using calibrated reconstruction method for both images, image1 and image2, three different algorithms of extracting features and features matching, named BM, SIFT, and CD, produce 3D models shown in Fig. 3. As illustrated in Fig. 4, the more efficient 3D model is achieved using SIFT algorithm. Applying different fusion methods of addition, average, PCA, and DWT, between SIFT-based model and CD-based model, four 3D models are shown in Fig. 5. As illustrated in Fig. 6, the more efficient 3D model is the PCA-based 3D model.

For uncalibrated reconstruction technique, applying fusion using PCA technique can produce a more accurate 3D model. As shown in Fig. 7, PCA-based calibrated 3D model is more accurate than the PCA-based uncalibrated 3D model, but it takes longer time. The resulted 3D models are used as targets for an artificial neural network. As in Fig. 8, the minimum error value is achieved for the proposed PCA-based 3D model.

5 Conclusions

This paper introduced a 3D object recognition approach using an artificial neural network. Efficient recognition system can be achieved by building accurate and complete 3D models. Different 3D reconstruction techniques can be used to produce 3D models. Different data fusion techniques can be applied to the produced 3D models. Experiments showed that 3D models built from performing calibrated reconstruction technique achieve better results than from uncalibrated one. It can be noticed that applying fusion between different 3D models can produce a more accurate result. The resulted 3D models are used as targets for a neural network, while different images that contain the object are used as inputs. The minimum recognition error value has been achieved for the 3D model produced by fusion based on PCA of 3D models produced from SIFT and CD.

References

1. Andrew EJ, Martial H (1999) Using spin images for efficient object recognition in cluttered 3d scenes. IEEE Trans Pattern Anal Mach Intell 21(5):433–449
2. Volker A, Jörg W (1995) Non-linear gaussian filters performing edge preserving diffusion. In: Mustererkennung, Informatik aktuell. Springer, pp 538–545
3. Olga M (2009) Image pre-processing tool. Kragujev J Math 32:97–107
4. Georgia K, Ellie KS, Andreas G, Anastasios D (2015) Hdr imaging for feature detection on detailed architectural scenes. In: 3D virtual reconstruction and visualization of complex architectures, vol XL-5/W4, pp 532–539
5. Theo M, Luc VG, Maarten V (2010) 3d reconstruction from multiple images part 1: principles. Comput Graph Vis 4(4):287–404
6. Richard IH, Peter S (1997) Triangulation. Comput Vis Image Underst 68(2):146–157
7. David GL (2004) Distinctive image features from scale-invariant keypoints. Int J Comput Vis 60(2):91–110
8. David GL (1999) Object recognition from local scale-invariant features. In: International conference on computer vision, vol 2. IEEE, pp 1150–1157

9. Tarek G, Alaa T, Abdelhameed I, Vǎclav S, Aboul Ella H (2015) Human thermal face recognition based on random linear oracle (rlo) ensembles. In: International conference on intelligent networking and collaborative systems. IEEE, pp 91–98
10. Kristian A, Wilfried K, Martin H, Andreas S (2009) Flexible hardware-based stereo matching. EURASIP J Embed Syst 2008:386059
11. Alexander H, Todd W (2010) Methods for enhancing corner detection
12. Demetri T, Andrew W, Michael K (1988) Symmetry-seeking models and 3d object reconstruction. Int J Comput Vis 1(3):211–221
13. Ted S, Luis AR (2003) 3D Reconstruction from two 2D images, ECE 533
14. Liang-Chia C, Hoang HH, Xuan-Loc N, Hsiao-Wen W (2013) Novel 3-d object recognition methodology employing a curvature-based histogram. Int J Adv Rob Syst 10(7):1–10
15. Jie T, Stephen M, Arjun S, Pieter A (2012) A textured object recognition pipeline for color and depth image data. In: International conference on robotics and automation. IEEE, pp 3467–3474
16. Kevin L, Liefeng B, Xiaofeng R, Dieter F (2011) Sparse distance learning for object recognition combining rgb and depth information. In: International conference on robotics and automation. IEEE, pp 4007–4013
17. Eunyoung K, Gerard M (2011) 3d object recognition in range images using visibility context. In: International conference intelligent robots and systems. IEEE, pp 3800–3807
18. Liefeng B, Xiaofeng R, Dieter F (2010) Kernel descriptors for visual recognition. Adv Neural Inf Process Syst 23:244–252
19. Kuk-Jin Y, Min-Gil S, Ji-Hyo L (2010) Recognizing 3d objects with 3d information from stereo vision. In: International conference on pattern recognition. IEEE, pp 4020–4023
20. Abdel karim B, Alaa FS, Mohammad SAB (2012) Feature based 3d object recognition using artificial neural networks. Int J Comput Appl 44(5):0975–8887
21. Rasool Reddy K, Phani Raja KMK, Madhava Rao Ch (2014) 3d reconstruction based on block matching and image pyramiding. Int J Sci Eng Res 5(10):1579–1586
22. Md Shafayat H, Ahmedullah A, Mohammad WR (2013) Rafi—a stereo vision based autonomous mobile area mapping robot with 16 dof. Int J Artif Intell Appl 4(1)
23. Naidu V, Raol J (2008) Pixel-level image fusion using wavelets and principal component analysis. Def Sci J 58(3):338–352
24. Abdelhameed I, Shoji T, Takahiko H (2009) Unsupervised material classification of printed circuit boards using dimension-reduced spectral information. In: International conference on machine vision applications, IAPR, pp 435–438
25. Alaa T, Abdelhameed I, Aboul Ella H, Gerald S (2015) Ear recognition using block-based principal component analysis and decision fusion. In: International conference on pattern recognition and machine intelligence. Springer, pp 246–254
26. Yufeng Z, Edward AE, Bruce CH (2004) An advanced image fusion algorithm based on wavelet transform: incorporation with PCA and morphological processing. In: Image processing: algorithms and systems III, vol 5298, SPIE
27. Vishal V, Suthar AC, Makwana YN, Shalini D (2010) Analysis of compressed image quality assessments. J Adv Eng Appl 225–229
28. Dumitru E, Christian S, Alexander T, Dragomir A (2014) Scalable object detection using deep neural networks. In: Conference on computer vision and pattern recognition. IEEE, pp 2155–2162

Digital Image Watermarking Through Encryption and DWT for Copyright Protection

Sarita P. Ambadekar, Jayshree Jain and Jayshree Khanapuri

Abstract Sharing of copyright documents, images, audios, and videos on Internet has become easier due to content authentication using digital watermarking. It also finds applications in the area of content protection, copyright management, and tamper detection. With the use of sophisticated signal/image processing algorithms, manipulations and duplications of audio, images, and videos are much easier. Hence, content authentication through encryption and resistance to general attacks such as noise, compression, and geometric has become an urgent and important issue. In this study, authors have proposed digital image watermarking technique based on discrete wavelet transform (DWT) and encryption. Watermark embedding and extraction algorithm using DWT coefficients, distance measurement, and encryption are demonstrated. DWT through multiresolution analysis provides the much needed simplicity in watermark embedding and extraction through watermark encryption. The technique results in PSNR greater than 50 dB and is resistance to noise, geometric, and compression attack. The proposed technique may be applied for copyright and content authentication applications.

Keywords Image watermarking · Discrete wavelet transform · Encryption
PSNR · Copyright protection

1 Introduction

The tremendous growth of high-speed LAN, WAN, MAN, and Internet technology have delivered means of new business, scientific, entertainment, and social prospects in the form of electronic broadcasting and marketing. The cost-effectiveness of sharing information and data in the form of digital documents, images, audios, and video

S. P. Ambadekar (✉) · J. Khanapuri
K. J. Somaiya Institute of Engineering and Information Technology, Mumbai, India
e-mail: sarita.ambadekar@somaiya.edu

J. Jain
IICE, Udaipur, India

S. Bhattacharyya et al. (eds.), *Recent Trends in Signal and Image Processing*,
Advances in Intelligent Systems and Computing 727,
https://doi.org/10.1007/978-981-10-8863-6_19

sequences by transmission over high-speed computer networks is greatly improved due to the advancement in Internet technology. Moreover transmission of digital information through Internet is very fast, low cost, and last but not least mostly without losing quality. Editing, copying, and tempering of data are easy because one can get access to the data through various means. Furthermore, a copy of a digital record is indistinguishable from the original record. Therefore, copyright protection and content authentication are becoming increasingly difficult tasks for digital data. Encryption/decryption techniques can be applied to restrict access to the data. However, encryption technique is unsuitable whenever the important data or information is decrypted and that can be manipulated and freely shared over the network. Digital image watermarking techniques are widely used for copyright protection and content authentication. However, continuous efforts are required to improve its performance due to new requirements and challenges such as multiple attacks and information sharing on social media Web sites [1–3]. A lot of information, authentic documents, government circulars, photographs, audios, and videos are shared on whats app and facebook. These original documents and images are subject to manipulations using sophisticated signal/image processing algorithms. Typically original images undergo various forms of manipulations such as cropping, geometric translation, contrast enhancement, and compression before being shared on Web sites or apps. Also when these images are shared and transmitted through communication networks are vulnerable to noise attack. Digital image watermarking plays important role in protection and authentication of images through watermark. Watermark can be an image embedded into the original image through watermarking algorithms. Yet watermark image can be visually invisible and recognizable when recovered from the original image even after being prone to multiple attacks [4–7]. It is possible to identify, remove, or change embedded watermark through sophisticated signal/image processing algorithms. Watermark being the identity of the content owner needs protection even after removed from the original document or image or video. Encryption provides additional security to the original watermark image in the event of unauthorized watermark extraction and manipulation. By the application of transforms such as DFT, DCT, and DWT, watermarking can be applied in the frequency domain [4–6]. Watermarking techniques using DCT are found to be more robust as compared to simple techniques applied in temporal domain [8]. Transform domain algorithms are robust against common signal and image processing operations like contrast enhancement, low-pass filtering, brightness adjustment, blurring [9–12]. However, their disadvantages are computationally expensive, difficult to implement and weak against geometric attacks like rotation, scaling, cropping [11]. DWT through multiresolution analysis provides the much-needed simplicity in watermark embedding and extraction through watermark encryption. DWT decomposes image with a normalized basis set. Thus, it can embed the watermark in any frequency band of a cover image [3]. In this study, authors have proposed digital image watermarking technique based on discrete wavelet transform (DWT) and encryption. Watermark embedding and extraction algorithm are devised and results are compared using watermarking parameters and its resistance to attack. The paper is organized as Sect. 2 describes the embedding and extraction algorithm, Sect. 3 illustrates results and concluded in Sect. 4.

2 Watermark Embedding Algorithm

Firstly, an image encryption algorithm based on row and column rotation through random number generator key k of the watermark image is performed. In this paper, we have used 90 × 90 pixels grayscale baboon image as watermark. Figure 1 shows the original watermark image and encrypted image.

Secondly, the original input image is decomposed using two-dimensional (2D) DWT to obtain the relevant scaled images with reduced size. Also, the encrypted watermark image is decomposed using 2D DWT to obtain decomposed scaled watermark images. Figure 2 shows the multiresolution decomposed images obtained after 2D DWT on original input and watermark image.

Thirdly, the pixel point at decomposed input image for embedding of the decomposed watermark image was identified based on Euclidean distance. More is the similarity between the input and watermark image, perceptibility of the input image does not change and increases the strength of watermark. Thus, it is more suitable for embedding watermark into the input image.

Fourthly, encrypted watermark was embedded into input image using (1) depending on the match between decomposed images of input and encrypted watermark image

$$y(i, j) = (1 - \text{alpha}) * i(i, j) + \text{alpha} * iw(i, j) \quad (1)$$

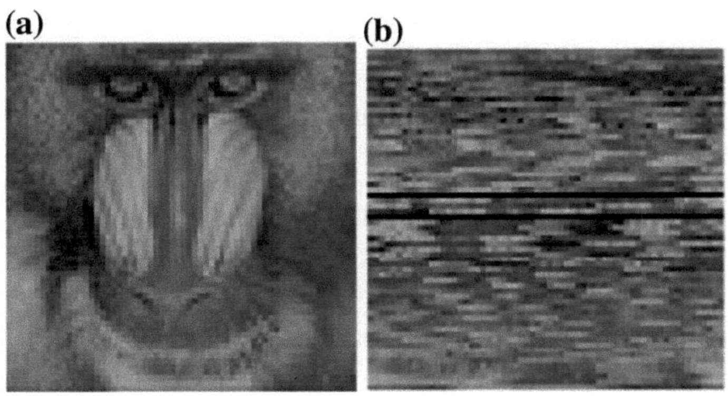

Fig. 1 **a** Original watermark image **b** encrypted watermark image

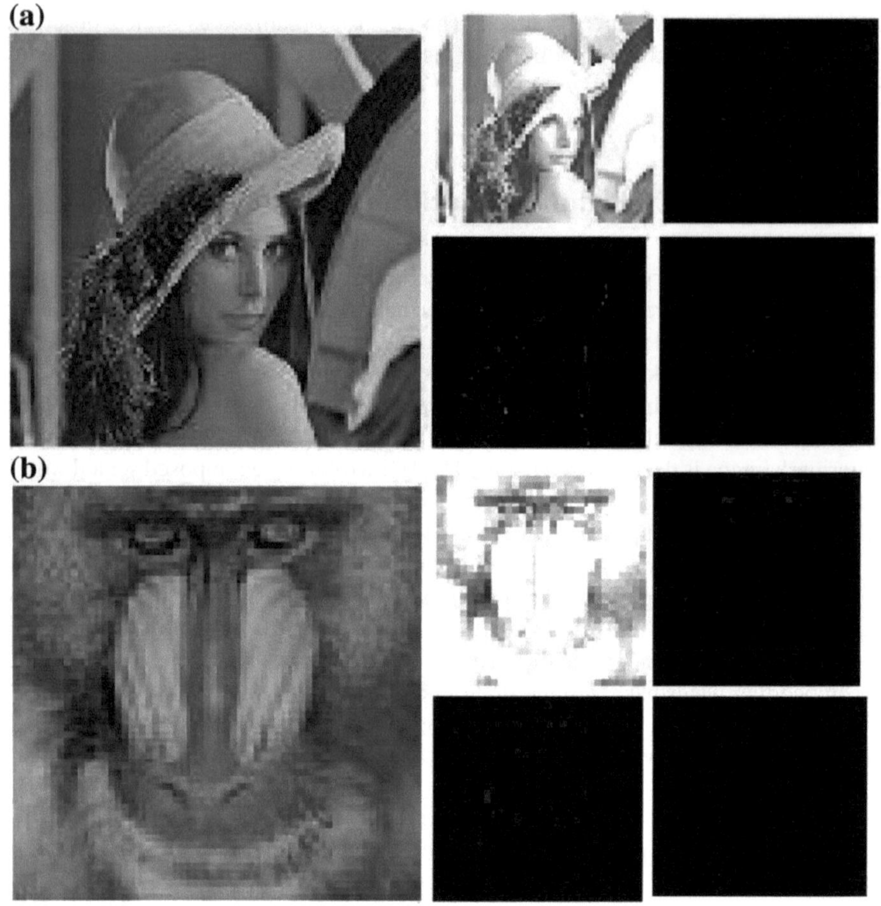

Fig. 2 Decomposition of image through 2D DWT **a** original input image **b** watermark image

where alpha is visibility coefficient, i(i, j) are DWT coefficients of respective decomposed input image, iw(i, j) are DWT coefficients of respective decomposed watermark image, and y(i, j) are the DWT coefficients of the watermark embedded output image.

The process for the watermark embedding algorithm is shown in Fig. 3.

3 Watermark Extraction and Detection Algorithm

The watermark extraction is exactly reverse procedure of watermark embedding. The algorithm presented in this paper is non-blind and therefore requires original input image and encryption key for watermark extraction and detection. The similarity between the original watermark image and extracted watermark image was measured

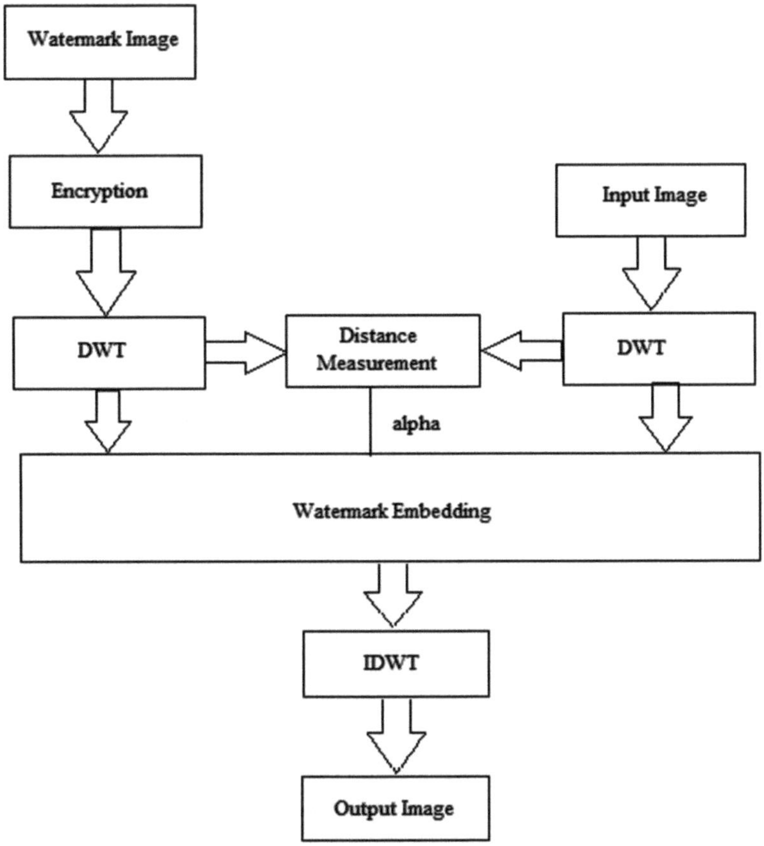

Fig. 3 Watermark embedding algorithm

using three parameters: mean square error (MSE), normalized correlation coefficient (CC), and peak signal to noise ratio (PSNR). In general, value of CC > 0.75 and PSNR > 30 dB is considered acceptable. Also, it is necessary to evaluate these watermarking parameters at various signal processing attacks. Watermark extraction algorithm is shown in Fig. 4.

4 Results and Discussion

Watermark embedding and extraction algorithm was implemented using MATLAB software and executed on intel i5 processor with 1 GB RAM and 3 GHz processing speed. In this paper, Lena image of size 228 × 228 and baboon image of size 90 × 90

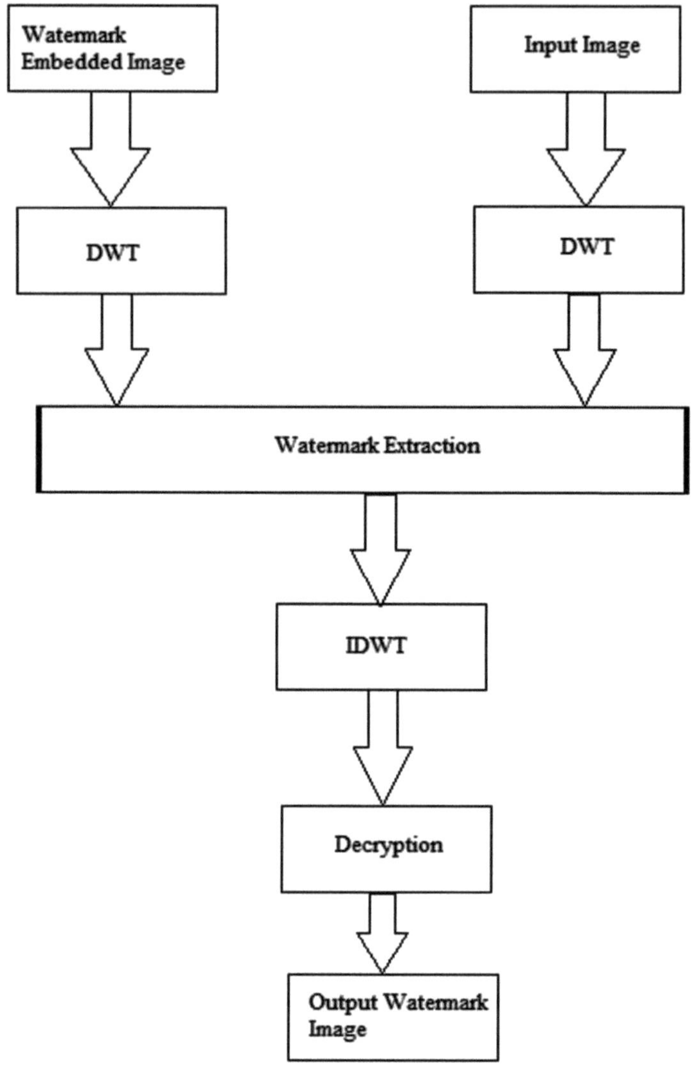

Fig. 4 Watermark extraction algorithm

were selected as input and watermark images, respectively. Figure 5 shows the original input image, watermark image, watermark embedded image, and extracted watermark image.

The performance of the presented algorithm is evaluated through three parameters MSE, CC, and PSNR. Also, the comparison of the experimentally obtained parameters was performed under two different conditions with and without attacks. Three general attacks such as salt-and-pepper noise, geometrical attack through rotation and JPEG compression attack were considered. In this experiment, salt-and-pepper

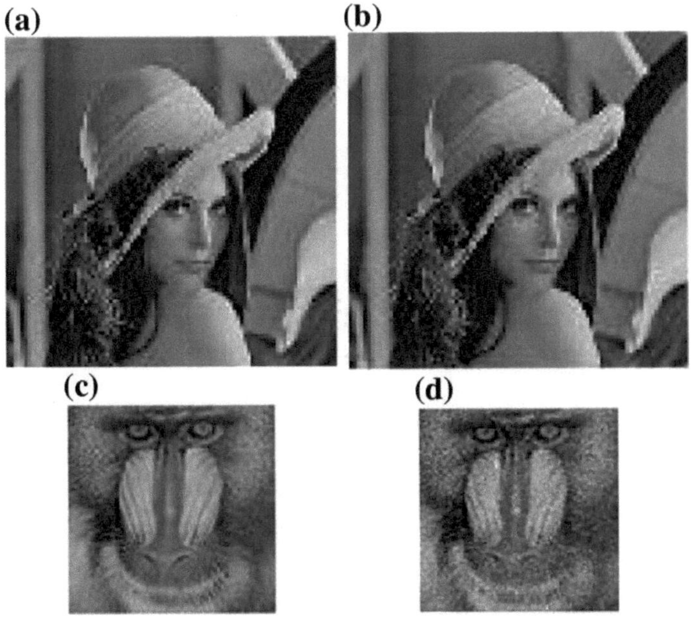

Fig. 5 **a** Original input image **b** watermark embedded image **c** original watermark image **d** extracted watermark image

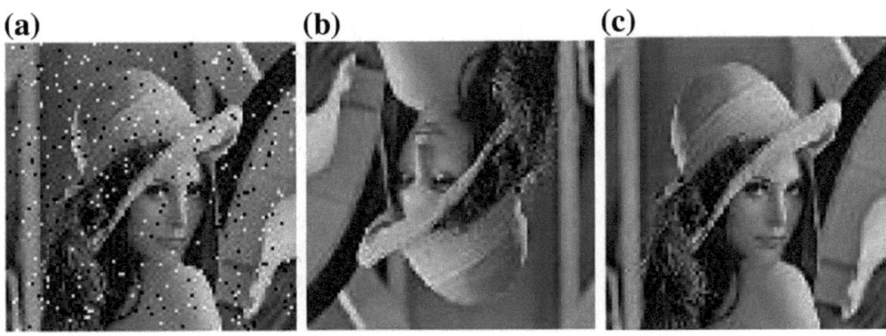

Fig. 6 Attacks on watermark embedded image **a** salt-and-pepper noise **b** rotation and **c** compression

noise of density 0.01, compression ratio of 2, and rotation of 90° was added into the watermark embedded image. Figure 6 shows the three types of attacks on watermark embedded image used for extraction algorithms. Table 1 shows the watermarking parameters evaluated between original watermark image and extracted watermark image at various attacks and without attacks. Comparison of PSNR with other methods through DWT is given in Table 2.

Obtain results validate the presented watermark embedding and extraction algorithm. DWT along with encryption provides better robustness to the algorithm for

Table 1 Watermarking parameters

Parameter	With attack			Without attack
	Noise	Geometric	Compression	
MSE	0.2053	0.2071	0.2033	0.2047
PSNR (dB)	55.006	54.96	55.04	55.01
CC	0.92	0.97	0.97	0.9749

Table 2 Comparison of PSNR (dB) values

Images	Proposed method	Vaidya et al. [2]	Peng et al. [13]
Lena	54.96	47.29	32.96

three general attacks. It has especially demonstrated better robustness against compression attack. Presented algorithm is simple to implement and provides some security to watermark through encryption key that can be suitable for applications such as Facebook and what's up that runs on android operating systems based devices.

5 Conclusion

In this paper, digital image watermarking algorithm through DWT and encryption for most common application facebook and whats app is presented. Robustness of the algorithm for general attacks such as salt-and-pepper noise, rotation, and compression is demonstrated. The presented algorithm is more suitable since simple convolution technique can be employed for computation of DWT, encryption with simple row and column rotation and pseudo-random number generator can be easily implemented on any device operating on android operating system. Further, robustness against geometrical distortions is focus of research in the implementation of watermarking algorithms.

References

1. Huang X, Zhao S (2012) An adaptive digital image watermarking algorithm based on morphological Haar wavelet transform. In: International conference on solid state devices and materials science, vol 25. Elsevier, Physics Procedia, pp 568–575
2. Vaidya P et al (2015) Adaptive digital watermarking for copyright protection of digital images in wavelet domain. In: 2nd international symposium on computer vision & internet, vol 58. Elsevier, Procedia Computer Science, pp 233–240
3. Chen L, Zhao J (2015) Adaptive digital watermarking using RDWT and SVD. In: IEEE international symposium on haptic, audio and visual environments and games (HAVE)
4. Roldan LR, Hernández MC, Chao J, Miyatake MN, Meana HP (2016) Watermarking-based color image authentication with detection and recovery capability. IEEE Lat Am Trans 14(2):1050–1057

5. Roy A, Maiti AK, Ghosh K (2015) A perception based color image adaptive watermarking scheme in YCbCr space. In: 2nd IEEE international conference on signal processing and integrated networks (SPIN)
6. Yadav N, Singh K (2015) Transform domain robust image-adaptive watermarking: prevalent techniques and their evaluation. In: IEEE international conference on computing, communication and automation
7. Shukla D, Tiwari N, Dubey D (2016) Survey on digital watermarking techniques. Int J Sig Process Image Process Pattern Recogn 9(1):239–244
8. Tao H et al (2014) Robust image watermarking theories and techniques: a review. J Appl Res Technol 12:122–138
9. Maity HK, Maity SP (2015) Multiple predictors based RW scheme with adaptive image partitioning. In: IEEE international conference on advances in computing, communications and informatics
10. Pushpa Mala S et al (2015) Digital image watermarking techniques: a review. Int J Comput Sci Secur 9(3):140–156
11. Maiorana E et al (2016) High-capacity watermarking of high dynamic range images. EURASIP J Image Video Process. Springer
12. Jagadeesh B et al (2015) Fuzzy inference system based robust digital image watermarking technique using discrete cosine transform. In: International conference on information and communication technologies, vol 45. Elsevier, Procedia Computer Science, pp 1618–1625
13. Peng F et al (2012) Adaptive reversible data hiding scheme based on integer transform. Sig Process 92(1):54–62

Speech and Non-speech Audio Files Discrimination Extracting Textural and Acoustic Features

Ghazaala Yasmin and Asit K. Das

Abstract It is a very accessible job to discriminate speech and non-speech audio files without giving too much effort. So, providing a sort of intelligence to a machine is required for the purpose. This issue is still coming out as an area of interest among the researchers. The prime goal of this work is to introduce a novel scheme for the discrimination of speech with non-speech audio data. The proposed methodology has introduced a new approach for identifying these two categories extracting both textural and acoustic features. As all features are not so important for discriminating the audio files, so some popular feature selection methods are applied to determine the important features and finally classification algorithms are applied on the reduced audio dataset. The experimental result shows the importance of the extracted features for the application.

Keywords Speech and non-speech audio discrimination · Textural features
Acoustic features · Feature extraction · Feature selection · Classification

1 Introduction

Real-world aura is permeated with extensive number of sounds. Those are broadly categorized into two sections: one is speech and the other is non-speech. Non-speech sounds are the general environmental sounds such as music, phone ring, thunder storm, raining animal sound. Speech sounds are the sounds delivered by human being containing meaningful words. Human brain has enough intelligence to figure

G. Yasmin (✉)
Department of Computer Science and Engineering, St Thomas' College of Engineering and Technology, Kolkata, India
e-mail: me.ghazaalayasmin@gmail.com

A. K. Das
Department of Computer Science and Technology, Indian Institute of Engineering Science and Technology, Shibpur, India
e-mail: akdas@cs.iiests.ac.in

S. Bhattacharyya et al. (eds.), *Recent Trends in Signal and Image Processing*, Advances in Intelligent Systems and Computing 727,
https://doi.org/10.1007/978-981-10-8863-6_20

out whether the sound is a speech or non-speech immaterial the person knows the language of the delivered speech sound or not. Robots and other speech equipment are getting benefited from this technology. Research is going on to resolve issue in a better way. This requisite has motivated us to introduce a new approach for differentiating speech and non-speech with a new set of features.

Speech and non-speech discrimination is the fundamental task of speech processing. Speech processing is very much discussed and researched area of nowadays because of its vast application area. Starting from security field to audio archive management, speech processing is found to be used. Once audio data is categorized into two broad areas—speech and non-speech further research can be carried out into each area. As a consequence, a large number of researchers are engaged in this domain to develop a robust and efficient feature set for the purpose of discriminating speech and non-speech. Thornton et al. [1] have worked with sensorimotor activity for speech and non-speech discrimination. They worked with 64-channel EEG data. They have performed time–frequency analysis of mu components. Grondin and Michaud [2] have worked with mobile robots. They have differentiated speech and non-speech for mobile robots by taking the help of pitch-based features. Rhythm-based work has been performed by Hiroya et al. [3]. Fuchs et al. [4] have also discriminated speech and non-speech applying electromyography (EMG). Their work was based on electro-larynx speech. Thambi et al. [5] have used frequency domain, time domain, and cepstral domain features. Several decision trees are applied in their work but among them random forest produces best result. Alexanderson et al. [6] have also distinguished speech and non-speech. They worked with motion captured by gestures in dialogue. Bowers et al. [7] have considered rhythm-based approach for discrimination of speech and non-speech. Rogers et al. [8] have considered theta-burst stimulation of motor cortex aiming to differentiation of speech and non-speech. Tremblay et al. [9] have performed discrimination of speech and non-speech in supratemporal plane. Oonishi et al. [10] have adopted a noise robust approach for speech and non-speech differentiation. They have considered GMM too in their approach. Izzad et al. [11] have identified speech and non-speech from spontaneous speech spoken in Malay language. They have decorated their feature vector based on energy, zero crossing rate (ZCR), and fundamental frequency (F0). A combined approach based on behavioral and event driven was followed by Reiche et al. [12] for measuring speech and non-speech deviations. Desplanques and Martens [13] have segmented speech and non-speech in multilingual TV broadcast. Elizalde and Friedland [14] have also discriminated speech and non-speech. They have worked with consumer-produced videos. They have discussed three approaches for this purpose—GMM, SVM, and NN based. K-NN algorithm has been used for classification of speech and non-speech audio by Priya et al. [15]. Bunton [16] has reviewed the extant studies related with oromotor non-speech activities to speech generation. An unsupervised approach was implemented by Maganti et al. [17] for detecting speech and non-speech. Their aim of the work was to recognize speech in meeting rooms. Ramírez et al. [18] have also discriminated speech and non-speech. Their decision statute was based on average LRT that is likelihood ratio test considering consecutive bispectrum speech features. Shin et al. [19] have also performed classification of speech and

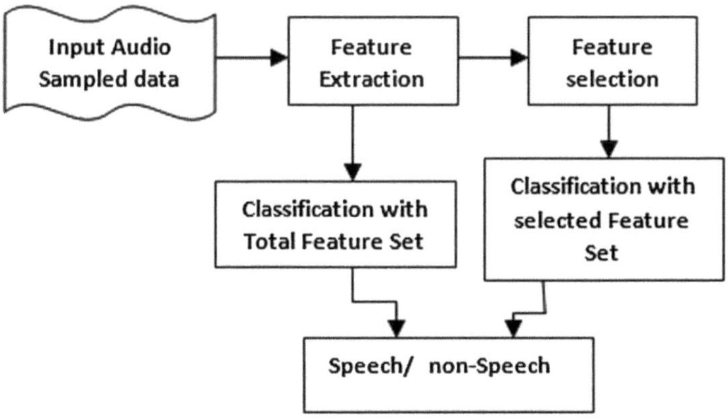

Fig. 1 Block diagram of a speech and non-speech discrimination system

non-speech files. They have suggested multiple features to improve the performance of endpoint detection. They have applied CART in their approach.

The rest of the paper is organized as follows: Sect. 2 describes the proposed work in details. Section 3 shows the experimental results of the proposed method followed by conclusion and future work in Sect. 4.

2 Methodology and Contribution

The prime goal of this methodology is to generate a unique set of features which can easily discriminate any type of speech and non-speech sounds. At first, the features have been extracted from sampled audio data. This whole feature set has been subjected to classification. Moreover, certain set of attributes in the feature set have been selected using some popular feature selection algorithms and the classification is being done based on these selected features. Figure 1 reflects the overall layout of the proposed methodology.

2.1 Feature Extraction

Speech signals differ from non-speech signal with respect to many measurable properties. This difference can be easily reflected in their acoustic features. This notion has inspired to select following textural and acoustic features.

(i) **Tempo-Based Features**

Every type of sound acquires its own speed. This can be measured by tempo. It has measurement of beats per minute. While discussing about speech and non-speech, it

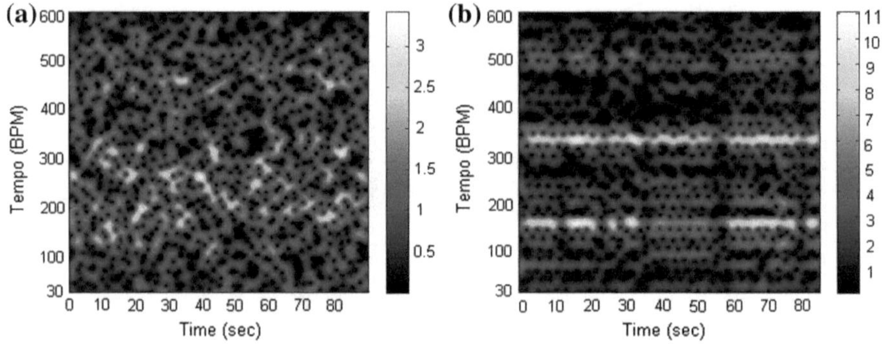

Fig. 2 Tempogram plot for speech (**a**) and non-speech (**b**)

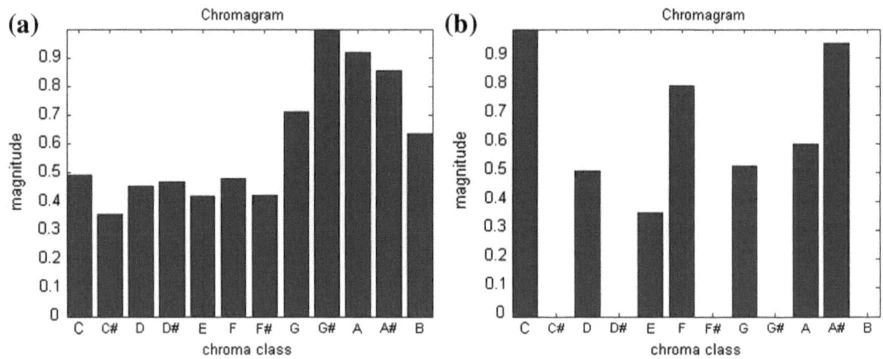

Fig. 3 Chromagram plot for speech (**a**) and non-speech (**b**)

is essential to keep in mind how the periodic occurrence of sound varies. Since this nature is clearly reflected by tempo-based feature, this feature has been picked up for the proposed feature set. The tempo-based feature has been extracted by computing the novelty curve from the sampled audio data.

The novelty curve has been broken into non-overlapped windows. Afterward, 30 Fourier coefficients have been derived from each tempo window. Finally, mean and standard deviation has been calculated considering all windows. This results a 60 dimension feature vector. Figures 2 and 3 show how the tempo is distributed over the time, whereas non-speech has been maintained its stable nature.

(ii) **Skewness**

Spectral skewness is a statistical feature, which derives the third-order moment of the spectral distribution. It is the measure of asymmetry present in a normal distribution with respect to its mean position. The suggested feature set is discriminati1ng speech and non-speech, where the signal pattern varies drastically. This feature shows discriminative result with respect to these two classes. The skewness sk is represented by Eq. (1).

$$sk = \frac{P(y-\mu)^3}{€^3} \tag{1}$$

where μ is the mean of sampled data y, € shows the standard deviation of y, and P(x) is the expected value of the quantity x.

(iii) **Chromagram-Based Features**

Chromagram features are the depiction for audio data where the complete spectrum is represented by twelve bins, indicating twelve different semitones (or chroma) of musical octave. It is already known that there is a basic dissimilarity persists between speech and non-speech with respect to the origin of the sound.

Being well known by this perception, it is also an obvious event that the chromagram representation of these genders will also vary. The intensity of each of the 12 individual musical chromagram of the octave, corresponding to each time frame is represented by the chromagram features. So, chromagram-based feature engenders 12 distinct values for an input audio file. In Fig. 3, it is clear that the chroma-based feature vector is acquiring its all chroma-based nature, and on the other hand, this value gets down to zero at some of its classes in case of non-speech.

(iv) **Zero Crossing Rate-Based Features**

Zero crossing rate (ZCR) can detect the presence of human speech in a signal as voice activity detection (VAD). It is a measure of how many times the audio signal crosses in the time domain. When the speech is delivered, it has been broken into multiple frames. For each frame, ZCR has been calculated, and henceforth, the final feature is taken into account by the mean and standard deviation of the values with respect to all frames.

Audio-clipped signal has been broken into R number of frames $\{x_i(n): 1 \leq i \leq R\}$. After that, for ith frame, zero crossing rates are calculated using Eqs. (2) and (3).

$$z_i = \sum_{n=1}^{m-1} sign\,[x_i(n-1) \ * \ x_i(n)] \tag{2}$$

where m specifies the number of sampled audio data present in the ith frame.

$$\boldsymbol{sign}[\boldsymbol{p}] = \begin{cases} 1, & if\ p > 0 \\ 0, & otherwise \end{cases} \tag{3}$$

In Fig. 4, it has been ascertained that the count for ZCR is kept on changing over the time slice, where non-speech signal has maintained its sustainability.

(v) **Co-occurrence Matrix Based Feature**

Generally mean and standard deviation are not enough and ultimate measurable parameter to determine the variation of feature values. The concept of co-occurrence matrix is the way to determine the precise changes in subjected feature from which the matrix is computed.

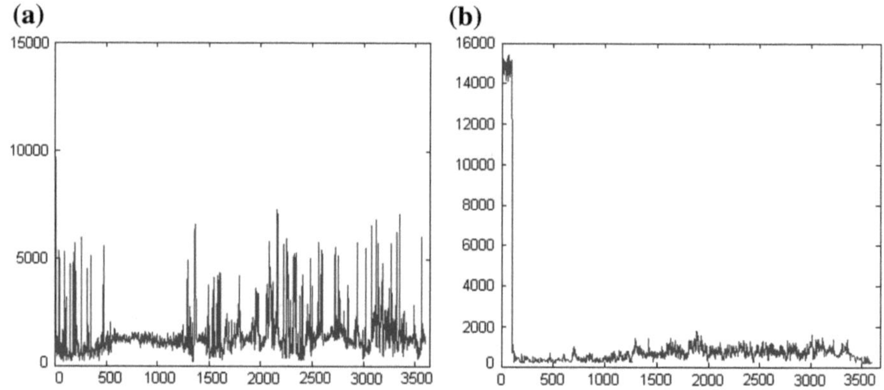

Fig. 4 ZCR plot for speech (**a**) and non-speech (**b**) sound

Let us assume, if y is the given feature vector and f(i) is the ith value of the feature vector and C(i, j) represents how often the value f(i) and f(i + 1) occurs in the entire feature vector.

Textural features are the computing parameter with respect to the statistical condition of the combination of values of the co-occurrence matrix. It has been appraised that textural features have emulated the fine nature of the features.

In the proposed work, co-occurrence matrix has been generated from the 12 coefficient values of chromagram-based features. Based on this co-occurrence matrix, the following texture features have been calculated.

(a) Inertia $= \sum(i-j)^2 M[i, j]$
(b) Entropy $= -\sum M[i, j] \log_2 M[i, j]$
(c) Energy $= \sum [M[i, j]]^2$
(d) Inverse difference $= \sum M[i, j]/|i - j|$ where $i \neq j$
(e) Correlation $= (1/\sigma_x \sigma_y) \sum \sum (i - \mu_x)(j - \mu_y) M[I, j]$

where, $M[i, j] =$ Value of co-occurrence matrix at position (i, j), $\mu_x = i \sum M[i, j]$, $\mu_y = j \sum M[i, j]$, $\sigma_x{}^2 = \sum(1 - \mu_x)^2 \sum M[i, j]$, $\sigma_y^2 = \sum(1 - \mu_y)^2 \sum M[i, j]$.

2.2 *Feature Selection and Classification*

The overall feature extracted from the audio sampled data contains 80 feature vectors. This contains 2 ZCR features, 60 tempo features, one skewness, 12 chromagram, and 5 textural features. There are many feature selection algorithms among which two popular feature selection techniques, namely correlation-based feature selection (CFS) and consistency-based feature selection (CON) algorithms have been applied to select the minimal set of important features. It has been observed that better classification accuracy is achieved after reducing the feature set. CFS algorithm

Table 1 Classification accuracy (in %)

Classifier type	Classifier name	Accuracy with all features	Accuracy with selected feature	
			CFS(14)	CON(5)
Bayes'	Bayes' net	92.19	**93.75**	93.75
	Naïve Bayes'	87.50	**89.06**	76.56
	Naïve Bayes' simple	88.50	**89.06**	81.45
Function	SVM	81.25	**84.36**	84.34
	Logistic	**98.44**	**98.44**	93.75
	MLP	99.1	**99.20**	60.94
Tree	J48	89.2	87.5	**92.19**
	Random forest	92.2	92.19	**93.75**
	Random tree	95.2	90.63	**95.31**
Rules	PART	89.2	89.06	**92.19**
	Decision table	92.1	93.75	**96.88**
	Ridor	92.3	87.7	**96.88**
Lazy	Kstar	96.8	**98.44**	95.31
	IB1	**98.8**	96.88	89.06
	IBk	**98.6**	96.88	89.06
Meta	Adabooster	**95.31**	**95.31**	**95.31**
	Bagging	89.06	89.06	**93.75**
	MCC (multiclass classifier)	**98.44**	**98.44**	93.75

reduced the total feature set into 14 features, whereas CON has selected only five features.

Classification is the process of dividing entire dataset into the distinct set of classes. It is a supervised learning technique that categorizes the dataset into different classes. There are many types of classification algorithms available in Waikato environment for knowledge analysis (Weka) tool [20], some of which are applied on the reduced dataset. The results are validated by 10-fold cross validation techniques, and classification accuracies are listed in Table 1.

3 Experimental Results

The collected sample audio data consists of 300 audio files having equal distribution for both categories. Noisy data has also been included in the dataset. The classification for training set and test set has been considered with tenfolds cross validation. Table 1 reflects the accuracy of different classifiers by taking all 80 features and reduced

Table 2 Comparative performance analysis of proposed work with other methods

Precedent method	Speech	Non-speech
Grondin and Michaud [2]	89.1	87.3
Hiroya et al. [3]	85.03	81.7
Thambi et al. [5]	90.4	89
Izzad et al. [11]	83	79.35

feature set obtained using CFS and CON. It is observed that, in case of Bayes' type, function type and Lazy type classifiers, CFS algorithm gives better accuracies and for tree type and rule-based classifiers, CON provides better classification accuracies. It can be concluded that the accuracy level has been improved by applying the feature selection algorithm. However, it has got impaired in some of the classification scheme. From this, it can be marked that feature selection technique is a significant task in the proffered methodology.

3.1 Comparative Analysis

For comparative analysis, the sampled dataset of the performed experiment has been utilized for implementation of the system proposed by Grondin and Michaud [2]. They have chosen a pitch-based feature for the classification of speech and non-speech sounds. Hiroya et al. [3] have adopted rhythm-based feature which has proven to be a good feature for this methodology. Thambi et al. [5] and Izzad et al. [11] have used frequency and time domain feature ZCR for their classification between speech and non-speech. From Table 2, it has been observed that the accuracy-level achieved better in proposed work compare to the other existing method.

4 Conclusions

Whenever the question comes with the security in audio data analysis, it is always the first task to identify whether the sound is speech or non-speech. A new approach of feature extraction has been propounded in the proposed work by introducing the textural feature based on chromagram feature vectors. It is aiming to apply the proposed system into other type of non-speech dataset such as computerized sound or some other digital sounds. The above experiment will be tested on some standard benchmark data. Conjointly the speech sound would be further explored for claiming speech password and subcategorizing of speech sound. It would be foresee to achieve the forthcoming goal with a promising outcome.

References

1. Thornton D, Harkrider AW, Jenson D, Saltuklaroglu T (2017) Sensorimotor activity measured via oscillations of EEG mu rhythms in speech and non-speech discrimination tasks with and without segmentation demands. Brain Lang
2. Grondin F, Michaud F (2016) Robust speech/non-speech discrimination based on pitch estimation for mobile robots. In: 2016 IEEE International Conference on robotics and automation (ICRA). IEEE, pp 1650–1655
3. Hiroya S, Jasmin K, Krishnan S, Lima C, Ostarek M, Boebinger D, Scott SK (2016) Speech rhythm measure of non-native speech using a statistical phoneme duration model. In: The 8th annual meeting of the society for the neurobiology of language
4. Fuchs AK, Amon C, Hagmüller M (2015) Speech/non-speech detection for electro-larynx speech using EMG. In: BIOSIGNALS, pp 138–144
5. Thambi SV, Sreekumar KT, Kumar CS, Raj PR (2014) Random forest algorithm for improving the performance of speech/non-speech detection. In: 2014 first international conference on computational systems and communications (ICCSC). IEEE, pp 28–32
6. Alexanderson S, Beskow J, House D (2014) Automatic speech/non-speech classification using gestures in dialogue. In: Swedish language technology conference
7. Bowers AL, Saltuklaroglu T, Harkrider A, Wilson M, Toner MA (2014) Dynamic modulation of shared sensory and motor cortical rhythms mediates speech and non-speech discrimination performance. Front Psychol 5
8. Rogers JC, Möttönen R, Boyles R, Watkins KE (2014) Discrimination of speech and non-speech sounds following theta-burst stimulation of the motor cortex. Front Psychol 5
9. Tremblay P, Baroni M, Hasson U (2013) Processing of speech and non-speech sounds in the supratemporal plane: auditory input preference does not predict sensitivity to statistical structure. Neuroimage 66:318–332
10. Oonishi T, Iwano K, Furui S (2013) A noise-robust speech recognition approach incorporating normalized speech/non-speech likelihood into hypothesis scores. Speech Commun 55(2):377–386
11. Izzad M, Jamil N, Bakar ZA (2013) Speech/non-speech detection in Malay language spontaneous speech. In: 2013 international conference on computing, management and telecommunications (ComManTel). IEEE, pp 219–224
12. Reiche M, Hartwigsen G, Widmann A, Saur D, Schröger E, Bendixen A (2013) Involuntary attentional capture by speech and non-speech deviations: A combined behavioral–event-related potential study. Brain Res 1490:153–160
13. Desplanques B, Martens JP (2013) Model-based speech/non-speech segmentation of a heterogeneous multilingual TV broadcast collection. In: 2013 international symposium on intelligent signal processing and communications systems (ISPACS). IEEE, pp 55–60
14. Elizalde B, Friedland G (2013) Lost in segmentation: three approaches for speech/non-speech detection in consumer-produced videos. In: 2013 IEEE international conference on multimedia and expo (ICME). IEEE, pp 1–6
15. Priya TL, Raajan NR, Raju N, Preethi P, Mathini S (2012) Speech and non-speech identification and classification using KNN algorithm. Proc Eng 38:952–958
16. Bunton K (2008) Speech versus nonspeech: Different tasks, different neural organization. In: Seminars in speech and language, vol 29, no 04. © Thieme Medical Publishers, pp 267–275
17. Maganti HK, Motlicek P, Gatica-Perez D (2007) Unsupervised speech/non-speech detection for automatic speech recognition in meeting rooms. In: IEEE international conference on acoustics, speech and signal processing, 2007. ICASSP 2007, vol 4. IEEE, pp IV-1037

18. Ramírez J, Górriz JM, Segura JC, Puntonet CG, Rubio AJ (2006) Speech/non-speech discrimination based on contextual information integrated bispectrum LRT. IEEE Signal Process Lett 13(8):497–500
19. Shin WH, Lee BS, Lee YK, Lee JS (2000) Speech/non-speech classification using multiple features for robust endpoint detection. In: 2000 IEEE international conference on acoustics, speech, and signal processing, 2000. ICASSP'00. Proceedings, vol 3. IEEE, pp 1399–1402
20. Markov Z, Russell I (2006) An introduction to the WEKA data mining system. ACM SIGCSE Bull 38(3):367–368

Speaker Recognition Using Occurrence Pattern of Speech Signal

Saptarshi Sengupta, Ghazaala Yasmin and Arijit Ghosal

Abstract Speaker recognition is a highly studied area in the field of speech processing. Its application domains are many ranging from the forensic sciences to telephone banking and intelligent voice-driven applications such as answering machines. The area of study of this paper is a sub-field of speaker recognition called speaker identification. A new approach for tackling this problem with the use of one of the most powerful features of audio signals i.e. MFCC is proposed in this paper. Our work also makes use of the concept of co-occurrence matrices and derives statistical measures from it which are incorporated into the proposed feature vector. Finally, we apply a classifier which correctly identifies the person based on their speech sample. The work proposed here is perhaps one of the first to make use of such an arrangement, and results show that it is a highly promising strategy.

Keywords MFCC · Co-occurrence matrix · Speaker recognition

1 Introduction

Speech is the medium which conveys not only the message being spoken but also reveals the information about the speaker. The chore which is easily being performed by the human such as face or speaker or emotion recognition proves difficult to emulate by computer. Speaker recognition stands out as an outperforming paradigm in

S. Sengupta · G. Yasmin
Department of Computer Science and Engineering, St. Thomas' College of Engineering and Technology, Kolkata, India
e-mail: ssengupta8@gmail.com

G. Yasmin
e-mail: me.ghazaalayasmin@gmail.com

A. Ghosal (✉)
Department of Information Technology, St. Thomas' College of Engineering and Technology, Kolkata, India
e-mail: ghosal.arijit@yahoo.com

S. Bhattacharyya et al. (eds.), *Recent Trends in Signal and Image Processing*, Advances in Intelligent Systems and Computing 727,
https://doi.org/10.1007/978-981-10-8863-6_21

the area of research. Speech signal holds noteworthy information about the speaker's identification. Moreover, speech data acquires some features which are enough to discriminate different speakers. Human speech has unique features which differentiate one person from other. Consequently, the imposition of technology claims the security in voice pathological and other forensic laboratory. This certainty has motivated to propose an efficient system to recognize speakers. Speaker recognition system has been proved to be a promising system for identifying telephone customers. Several organizations deployed this system as the authenticating tool [1]. The objective of this work is to nominate a unique idea to identify different speakers from a set from audio speech.

1.1 Related Work

Past studies have discriminated the speaker voice using Gaussian mixture model [1]. Dudeja and Kharbanda [2] have introduced text-independent speaker recognition using Mel Frequency Cepstral Coefficient (MFCC). The concept of MFCC has also been taken into account with text-dependent approach [3]. Revathi et al. [4] have adopted the concept of clustering for speaker identification. Reynolds et al. [5] have worked on speaker verification. Their approach was based on Gaussian mixture model (GMM). Kua et al. [6] have also worked with speaker recognition. They have worked with Spectral Centroid Frequency (SCF) as well as Spectral Centroid Magnitude (SCM). Doddington [7] has recognized speaker depending on idiolectal differences existing between any two speakers. Vector quantization approach for speaker recognition has been adopted by Kinha and Vig [8]. MFCC was also used in their work. Paul and Parekh [9] have worked with isolated words and neural networks for speech recognition. Zero Crossing Rate (ZCR) also has been used by them to achieve their goal. Otero [10] has worked with speaker segmentation. Campbell [11] has provided a tutorial on speaker recognition. The concept of linear predicting coding has also been introduced as a discriminative feature by Atame et al. [12]. Mermelstein [13] has worked with psychological representation of speech sounds. Haralick [14] has dealt with structural and statistical approaches towards speakers. Different textural features of image processing field have been discussed in that work. Lartillot et al. [15] have developed a toolbox in MATLAB to serve the purpose of retrieval of music information. This toolbox is named as 'MIRToolbox'. All the proposed system for speaker recognition primarily has to serve two phases. First is to choose suitable architecture for the proposed system, and second is to adopt efficient feature. The suggested approach has been expected to arise an easy and able way to recognize speaker by speech data. Perrachione [16] has tried to recognize speakers for different languages.

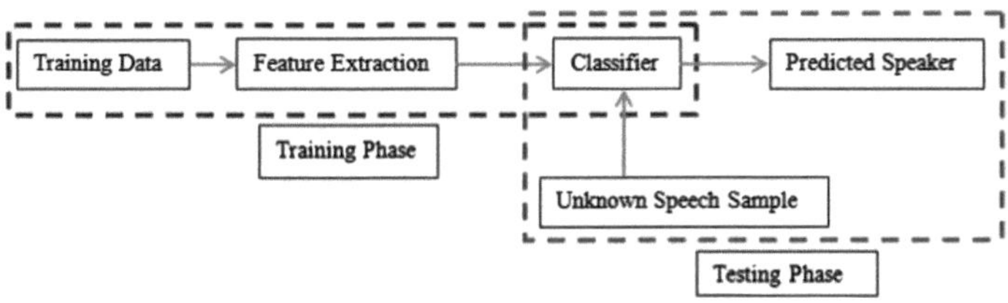

Fig. 1 Block diagram of a speaker identification system

2 Proposed Methodology

A speaker identification system as the name suggests aims to answer the question 'Whose voice is being analysed?'. The models in this area of work operate generally on a three-stage architecture, namely involving the phases (i) feature extraction from the training data, (ii) classifier training and (iii) identification of the speaker from their testing sample. Figure 1 outlines the entire idea.

2.1 Foundational Ideas Behind MFCC

The proposed work aims to make use of Mel Frequency Cepstral Coefficients [13] (MFCCs) to determine the speaker whose speech sample is being analysed. These features are of utmost importance as they help in capturing salient or prominent information from a speech signal. MFCCs are based on psychoacoustic modelling of sound. It strives to capture an individual's vocal tract properties through the envelope of the short time power spectrum. These features are extremely robust and relatively computationally inexpensive in calculating and as such have garnered much popularity in the field of Automatic Speech Recognition (ASR) in recent years. Sound or speech of any kind produced by human beings is defined by the shape of their vocal tract as well as tongue, teeth, and larynx among others. MFCC is a kind of feature which helps detect this shape and thus the person associated with it which forms the fundamental basis of its construction, as each person has a unique auditory system. This feature generates values (coefficients) which are produced after passing a speech signal through a series of steps which closely mimics the way a human ear comprehends sound.

2.2 MFCC Extraction Procedure

The steps involved in the extraction of such cepstral coefficients are enumerated below and described diagrammatically in Fig. 2.

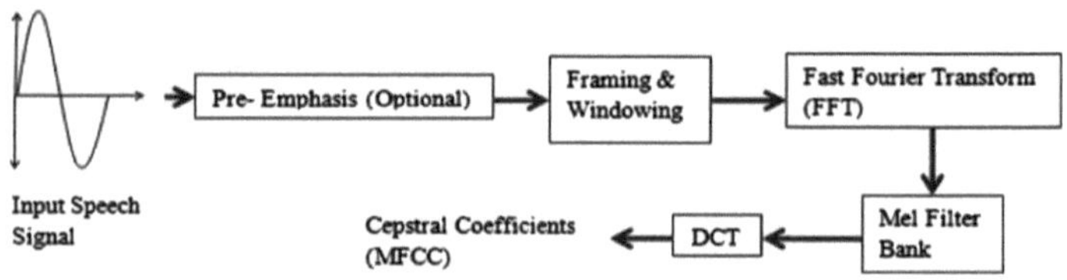

Fig. 2 MFCC extraction process

- Pre-emphasis: This is actually an optional phase in the extraction process. In this stage, all the frequencies from the input speech signal are analysed and the amplitude (energy) of the high frequency bands is increased, while the amplitude of low frequency bands is decreased because it was found out through experiments that higher frequencies are more useful for distinguishing between different signals than low frequencies. Pre-emphasis is useful for dealing with DC offset which in turn can improve the performance of energy-based voice activity detection (VAD). It is known that the energy of lower frequency signals degrades over time while being transmitted and as such, pre-emphasis can help in boosting the signal to a high frequency. Even with all these advantages, it is still an optional phase because of the efficiency of computing power which is brought about by modern times. Pre-emphasis was used in older models for feature extraction owing to the lack of computing power. This stage can be skipped over because it is accounted for in later stages with channel normalization techniques like cepstral mean normalization.
- Framing and Windowing: In this phase, the input signal is broken up into small segments of duration 20–30 ms called frames. The reason for breaking up the signal into frames is because of the time-varying nature of signal properties. A signal of any kind fluctuates in property as time goes by. Thus, analysing it in its entirety would not be correct and as such, it is divided into smaller parts (frames) assuming that in all fairness the properties remain stationary.
  After framing the signal, it essentially becomes a discrete set of frames unlinked from each other. In order to remove the discontinuities at the start and end of a frame, each frame is multiplied by a quantity known as a window function. After windowing, the spectral distortion in the signal gets minimized.
- Fast Fourier Transform (FFT): So far, the signal which was being dealt with was present in the time domain. In order to extract the cepstral coefficients, it must be converted into its frequency domain representation which is exactly what the Fourier transformation does. The output of this stage is a frequency spectrum or periodogram which determines what frequencies are present in the given signal. The periodogram acts very much like the human auditory system (particularly the cochlea) because depending on the frequencies present in the perceived sound certain spots on the cochlea resonate which in turn activates different nerve endings informing the brain which frequencies are present. A periodogram obtained from FFT aims to replicate this idea as it determines which frequencies compose the given signal and how much power is associated with each frequency.

- Mel Filter Bank: The periodogram produced from the previous stage contains all the frequencies present in the supplied signal. But all of these frequencies are not relevant enough to be studied because at very high frequency values, auditory perception becomes blurred. Because of this reason, chunks of the periodogram are selected to get an idea of how much energy exists in the different frequency bands. This is the work of the Mel filter bank. The Mel scale is used to define the boundaries of each frequency band, i.e. how wide or narrow to make them, etc.
- Discrete Cosine Transformation (DCT): Once the filter-bank energies have been obtained, we take the logarithm of them because this is again due to human perception of sound, as we do not perceive sound on a linear scale. At last, DCT of the filter-bank energies is calculated. The main reason for computing DCTs is that the energies are highly correlated and it is desired that they be uncorrelated so that they may be used with classification algorithms. A total of 26 DCT coefficients are obtained from the filter-bank energies, and only the first 12 or 13 are retained. This is because the higher coefficients represent fast and sudden changes in the filter-bank energies and if used, the system performance gets degraded. Ultimately, these 12 or 13 coefficients are called Mel Frequency Cepstral Coefficients.

2.3 Co-occurrence Matrix

Co-occurrence matrices are a well-defined tool finding applications in image processing. An image is basically a collection of pixels, and each pixel has neighbouring pixels in generally eight directions, viz. North, South, East, etc., (spatial relationships). The idea here is to determine how many times a particular pixel pair occurs together in an image and represents this frequency in the form of a matrix (co-occurrence). Thus, the cells of a co-occurrence matrix are populated with the frequencies of occurrence of a particular pixel pair specified for a certain direction. This means that the pixel pair (1, 2) and (2, 1) are different. A sample pixel matrix representation for an image along with its corresponding co-occurrence matrix is shown in Fig. 3. It should be noted that we consider only the East spatial relationship between the pixels when computing the co-occurrence matrix for our example image. The co-occurrence matrix is also called Grey Level Co-occurrence Matrix (GLCM) because an image in its simplest form is a matrix of greyscale values lying in the range [0, 255]. This is owing to the fact that each pixel (intensity value) is internally represented as an 8-bit number and 28 is equal to 256. Zero on this scale indicates absolute white and 255 absolute darkness (black).

From a co-occurrence matrix, several statistical features (called Haralick [15] texture features) can be obtained. If M is the co-occurrence matrix and $1 \leq i \leq$ rows (M) and $1 \leq j \leq$ columns (M) where rows (M) are the number of rows in M and similarly columns (M) are the number of columns in M, then the features are defined as follows:

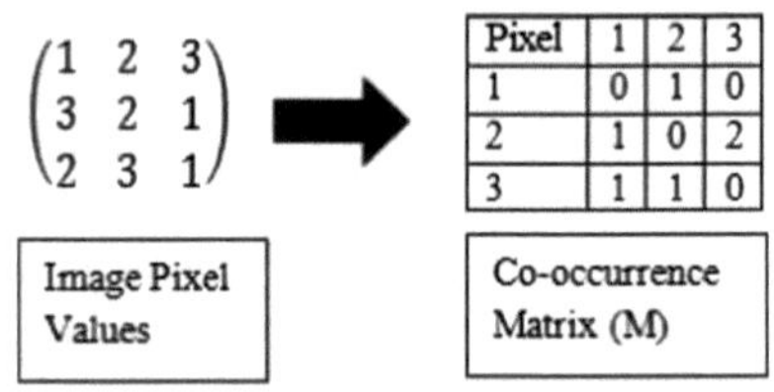

Fig. 3 Co-occurrence matrix of an image

$$\text{Contrast} = \sum_{i,j} |i - j|^2 M[i][j]. \text{ Contrast is also known as inertia or variance.} \tag{1}$$

$$\text{Correlation} = \sum_{i,j} (i - \mu_x)\left(j - \mu_y\right) M[i][j]/\sigma_x \sigma_y \tag{2}$$

$$\text{Energy} = \sum_{i,j} M[i][j]^2 \tag{3}$$

$$\text{Homogeneity} = \sum_{i,j} M[i][j]/1 + |i - j| \tag{4}$$

$$\text{Entropy} = -\sum_{i,j} M[i][j] \log_2 M[i][j] \tag{5}$$

where,

- $\mu_x = \sum_i i \sum_j M[i][j]$
- $\mu_y = \sum_j j \sum_i M[i][j]$
- $\sigma_x = \sum_i (1 - \mu_x)^2 \sum_j M[i][j]$
- $\sigma_y = \sum_j \left(1 - \mu_y\right)^2 \sum_i M[i][j]$

2.4 Experimental Set-up and Implementation Details

A custom database was used in our work which contained speech recordings from various speakers, as obtaining a standardized speech database is a difficult task. The files were created in the '.wav' format, owing to its uncompressed signal nature, using suitable text-to-speech (TTS) software. The reason for choosing TTS software is that it was desirable to have at least 10 recordings from an individual speaker and naturally it would have been quite a laborious task to ask our speakers to record 10 samples each. We took a total of 10 speakers, 5 male and 5 female, and produced 10 recordings for each, using standardized text, of duration of 1 min and 30 s so as to create a more robust model for classification with large training data. Thus, our database was created and subsequently the feature vectors for each could be extracted.

As mentioned before, the speech files needed to be appropriately framed in order to extract the cepstral coefficients. A 50 ms frame size was chosen with half overlap (the succeeding frame beginning at the middle of the former) between adjacent frames.

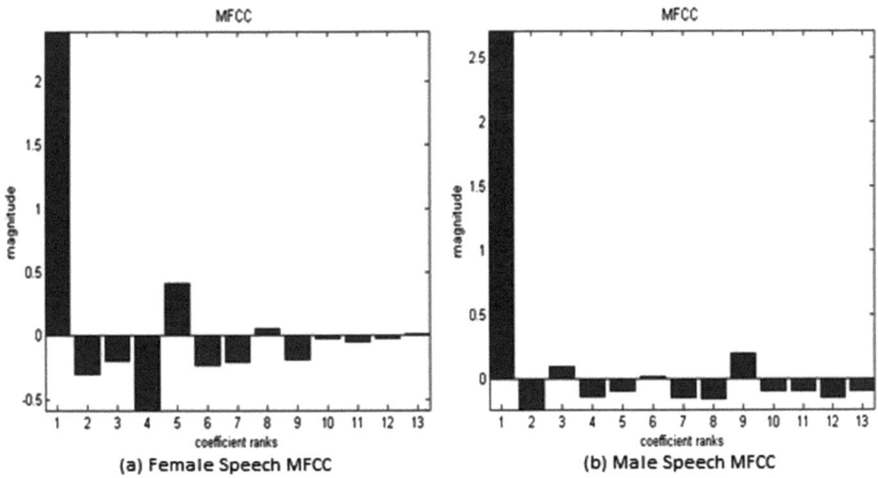

Fig. 4 MFCC plots for female and male speech signals

Once framed, a set of 13 Mel cepstral coefficients were derived for each frame giving in total a matrix of dimension 13 $\times$ F where F is the total number of frames in the signal. Following this, a co-occurrence matrix was obtained from the cepstral matrix and the five statistical measures described in [14] were calculated using it. Finally, from each frame the mean of each coefficient was calculated resulting in a final 13-dimensional cepstral vector for the entire signal. This means that the first coefficient in this vector is the mean of the first coefficient taken from all the frames, etc. As such, an 18-dimensional feature vector was produced for each file where the cepstral coefficients for the entire signal occupied the first 13 dimensions while the remaining 5 were filled up by the statistical features obtained from the co-occurrence matrix of the cepstral coefficients. We calculate the co-occurrence matrix of the MFCC features for obtaining a better understanding of the signal properties. In order to extract the Mel coefficients, we took help of the Music Information Retrieval (MIR) toolbox [15].

Once the feature set was produced, a classifier needed to be trained on the data. In our experiments, the k-NN (Nearest Neighbour) classifier was chosen. After the training phase, an unknown speech sample was provided to it and the response was checked to see whether it was predicting the correct speaker or not.

3 Results and Analysis

A plot for MFCCs of both male and female speech is shown in Fig. 4, and Table 1 describes the prediction results from our experiments.

Table 1 Prediction results

Trial number	True speaker	Predicted speaker	Correctness of prediction
1	M1	M1	1
2	M2	M2	1
3	M3	M3	1
4	M4	M4	1
5	M5	M4	0
6	F1	F1	1
7	F2	F2	1
8	F3	F3	1
9	F4	F4	1
10	F5	F5	1
		Total	9

Here 'Mi' stands for the ith Male and 'Fi' for the ith Female speaker ($1 \leq i \leq 5$). The correctness of prediction column is defined as a binary result with 0 meaning an incorrect prediction and 1 meaning a correct one. From Table 1, it is seen that out of 10 speakers, our system was able to correctly identify 9 of them, thus indicating an accuracy of 9/10 or 90%. The k-NN classifier was kept in its default settings, and still such a high level of classification accuracy was obtained highlighting the improvements brought about by the approach. We wanted to check how low the threshold of error was in our system which meant intentionally supplying an unknown speech sample not belonging to any of the speakers in the database and measuring the correlation between the predicted speakers feature vector and the supplied speakers feature vector. Under such a set-up, we observed a 0.97 or 97% correlation (Pearson) between the unknown sample and the predicted speaker. Such a high correlation indicates the fact that our system faces difficulty in distinguishing between different speakers whose voice patterns are highly similar to one other. This fact could also indicate that our system is not suitable for handling cases involving imposter or fraud detection. Keeping all of this in mind, credit must be given to the proposed feature system as it fails only in cases where there is an extremely high similarity (correlation) between an unknown sample and an existing speaker in the database, thus proving that the threshold of error in our model is competitively low.

3.1 Comparative Study

To judge the strength of proposed feature set to recognize speaker, it has been compared with other research works. Features proposed by different researchers have been applied in this data set. The comparison has been given in Table 2, and it indicates that proposed feature set performs better than other research works.

Table 2 Comparative analysis

Method	Classification accuracy (%)
Revathi, Ganapathy and Venkataramani	86.3
Kua, Thiruvaran, Nosratighods et al.	87.6

4 Conclusions and Future Direction

The field of speaker recognition is garnering much attention in modern times because of its vast spectrum of application areas ranging from forensic science to hands-free computing. To that end, research is being carried out extensively in order to improve the quality of recognizing a person from their speech sample. A new approach has been proposed in this paper describing how the task of speaker identification can be performed using the co-occurrence pattern of MFCC features. Even while using such a low-dimension feature vector, we are seeing high levels of prediction accuracy. But as can be understood from our experiments, there still remains a lot of room for improvement with the ultimate goal in mind to obtain recognition accuracy as high as possible. There have been numerous advents into speaker identification using MFCC as the primary feature for discrimination. But our work is perhaps the first to observe the recognition problem in the light of the co-occurrence pattern of MFCC.

Acknowledgements This work is partially supported by the facilities created under the open-source software AudioBookMaker 2.01.

Disclaimer

We the authors have obtained all ethical approvals from an appropriate ethical committee. The consent of the speakers has been taken, to use the data in this research. Neither the editors nor the publisher will be responsible for any misuse or misinterpretation of the data.

References

1. Reynolds DA (1995) Automatic speaker recognition using Gaussian mixture speaker models. Lincoln Lab J
2. Dudeja K, Kharbanda A (2015) Applications of digital signal processing to speech recognition. Int J Res 2(5):191–194
3. XU HH (2015) Text dependent speaker recognition study
4. Revathi A, Ganapathy R, Venkataramani Y (2009) Text independent speaker recognition and speaker independent speech recognition using iterative clustering approach. Int J Comput Sci Inf Technol (IJCSIT) 1(2):30–42
5. Reynolds DA, Quatieri TF, Dunn RB (2000) Speaker verification using adapted Gaussian mixture models. Digit Signal Process 10(1–3):19–41
6. Kua JMK et al (2010) Investigation of spectral centroid magnitude and frequency for speaker recognition. Odyssey 34–39
7. Doddington GR (2001) Speaker recognition based on idiolectal differences between speakers. Interspeech 2521–2524
8. Suraina K, Vig R (2015) A mfcc integrated vector quantization model for speaker recognition. Int J Comput Sci Mob Comput 4(5):294–400

9. Paul D, Parekh Ranjan (2011) Automated speech recognition of isolated words using neural networks. Int J Eng Sci Technol (IJEST) 3(6):4993–5000
10. Otero PL (2015) Improved strategies for speaker segmentation and emotional state detection
11. Campbell JP (1997) Speaker recognition: a tutorial. Proc IEEE 85(9):1437–1462
12. Atame S, Shanthi Therese S, Madhuri G (2015) A survey on: continuous voice recognition techniques. Int J Emerg Trends Technol Comput Sci (IJETTCS) 4(3):37–41
13. Mermelstein P (1976) Distance measures for speech recognition, psychological and instrumental. Pattern Recog Artif Intell 116:374–388
14. Haralick RM (1979) Statistical and structural approaches to texture. Proc IEEE 67(5):786–804
15. Lartillot O, Toiviainen P, Eerola T (2008) A matlab toolbox for music information retrieval. In: Data analysis, machine learning and applications, pp 261–268
16. Perrachione TK (2017) Speaker recognition across languages. Oxford University Press

Author Index

S. Bhattacharyya et al. (eds.), *Recent Trends in Signal and Image Processing*,
Advances in Intelligent Systems and Computing 727,
https://doi.org/10.1007/978-981-10-8863-6

MIX
Papier aus verantwortungsvollen Quellen
Paper from responsible sources
FSC® C105338

If you have any concerns about our products,
you can contact us on
ProductSafety@springernature.com

In case Publisher is established outside the EU,
the EU authorized representative is:
Springer Nature Customer Service Center GmbH
Europaplatz 3, 69115 Heidelberg, Germany

Printed by Libri Plureos GmbH
in Hamburg, Germany